Rozanov   Introduction to Random Processes

Yuriĭ A. Rozanov

Introduction to
Random Processes

Translated from the Russian
by Birgit Röthinger

Springer-Verlag
Berlin Heidelberg New York
London Paris Tokyo

Yuriĭ A. Rozanov
Steklov Mathematical Institute
ul. Vavilova 42, 117333 Moscow, USSR

Birgit Röthinger
Franz-Knauff-Str. 17, D-6900 Heidelberg

Title of the Russian original edition:
Vvedenie v teoriiu sluchaĭnykh protsessov
Publisher Nauka, Moscow 1982

This volume is part of the *Springer Series in Soviet Mathematics*
Advisers: L. D. Faddeev (Leningrad), R. V. Gamkrelidze (Moscow)

ISBN-13: 978-3-642-72719-1   e-ISBN-13: 978-3-642-72717-7
DOI: 10:1007/978-3-642-72717-7

Library of Congress Cataloging-in-Publication Data
Rozanov, ÎŬ. A. (ÎŬriĭ Anatol´evich), 1934-
Introduction to the Theory of Random Processes.
(Springer series in Soviet mathematics)
Translation of: Vvedenie v teoriiŭ sluchaĭnykh profsessov.
1. Stochastic processes. I. Title. II. Series.
QA274.R69313 1987 519.2 87-16305

Softcover reprint of hardcover 1st edition 1987

© Springer-Verlag Berlin Heidelberg 1987

Printing: Druckhaus Beltz, Hemsbach/Bergstr.
Bookbinding: J. Schäffer GmbH & Co. KG., Grünstadt
2141/3140-543210

Contents

Preface

Today, the theory of random processes represents a large field of mathematics with many different branches, and the task of choosing topics for a brief introduction to this theory is far from being simple.

This introduction to the theory of random processes uses mathematical models that are simple, but have some importance for applications. We consider different processes, whose development in time depends on some random factors. The fundamental problem can be briefly circumscribed in the following way: given some relatively simple characteristics of a process, compute the probability of another event which may be very complicated; or estimate a random variable which is related to the behaviour of the process. The models that we consider are chosen in such a way that it is possible to discuss the different methods of the theory of random processes by referring to these models.

The book starts with a treatment of homogeneous Markov processes with a countable number of states. The main topic is the ergodic theorem, the method of Kolmogorov's differential equations (Secs. 1-4) and the Brownian motion process, the connecting link being the transition from Kolmogorov's differential-difference equations for random walk to a limit diffusion equation (Sec. 5). The chapters that follow outline the foundations of stochastic analysis. Specifically, we treat random processes as curves in the space of random variables with the norm of quadratic mean (Secs. 7, 8). Then we show how random processes can be described by linear stochastic differential equations, and we explore their convergence behaviour, particularly their convergence towards processes that are stationary in the wide sense. In Secs. 12 and 13 we deal with the fundamentals of the spectral analysis of stationary processes. Finally, some special problems of estimation and filtration are discussed in Secs. 14 and 15. In Sec. 6, which stands somewhat apart from the rest, an attempt is made to apply direct probabilistic methods for sums of independent, identically distributed, variables to a multi-server system. As a complement, you may read Secs. 9-11 where nonlinear stochastic differential equations for diffusion processes are dealt with.

For supplementary study we would recommend the book by S. Karlin, "A first course in stochastic processes" (New York and London, Academic Press, 1966), which gives more attention to the topics explained in Secs. 1-6 of our book, especially to the examples from biology, genetics and from the multi-server system. We also recommend the book of A. D. Ventsel, "A course in the theory of stochastic processes" (Russian: Moscow, Nauka, 1975, English: New York, McGraw Hill, 1981).

We have included important material in the form of problems and hints.

The appendix at the end of the book briefly explains basics of probability theory. The reader can find a systematic treatise of the foundations of probability theory in the books of A. A. Borovkov, "Probability theory" (Moscow, Nauka, 1976) and A. N. Shiryayev, "Probability" (Russian: Moscow, Nauka, 1980, English: New York, Springer-Verlag, 1984).

Section 1
Random Processes with Discrete State Space
Examples

We consider the process of radioactive decay where radium **Ra** is converted into radon **Rn** over a period of time. In this process, at the time of disintegration, the **Ra** atom emits an α-particle (the nucleus of the helium atom **He**) and changes from **Ra** into **Rn**. It is well known that this process is a random process.

We suppose that each **Ra** atom is converted to a **Rn** atom in the course of time with some probability $F(t)$ dependent on t. More precisely, we talk about the distribution function $F(t)$ of the time of decay of a single **Ra** atom which has been observed from some initial time t_0, that is

$$(1.1) \qquad F(t) = P\{\tau \leqslant t\}, \quad t \geqslant 0,$$

where τ is the period between the time t_0 and the time of the transition **Ra** $\rightarrow$ **Rn**. We assume that the probability $P\{\tau \leqslant t\}$ of decay up to time t is always the same, independently of the time t_0 at which we chose one of the **Ra** atoms.

Accordingly, after observing a selected **Ra** atom from the time t_0 and under the condition $\tau > s$, at time $t_1 = t_0 + s$, we have the same **Ra** atom as before, and the probability of its disintegration during the following period of time t is $F(t)$. The probability of its remaining intact, i.e. the probability that $\tau > s + t$ with $\tau > s$, is $1 - F(t)$. We introduce the function

$$p(t) = 1 - F(t) = P\{\tau > t\}, \quad t \geqslant 0.$$

We can show that $p(t) = P\{\tau > t\}$ corresponds to the conditional probability of $\tau > s + t$ under the condition $\tau > s$:

$$P\{\tau > s + t \mid \tau > s\} = p(t) = P\{\tau > t\}.$$

Applying this for the probability that $\tau > s + t$, we get

$$p(s + t) = P\{\tau > s + t\}$$

$$= P\{\tau > s + t \mid \tau > s\} \cdot P\{\tau > s\} = p(t) \cdot p(s).$$

So we have for all $s, t \geqslant 0$

(1.2) $p(s + t) = p(s) \cdot p(t).$

This is well known in the analysis of functional equations and entails the function

(1.3) $p(t) = P\{\tau > t\} = e^{-\lambda t}, \quad t \geqslant 0.$

Of course, we are talking about a function $p(t)$ not identically equal to 0. We can easily deduce formula (1.3) from the functional equation (1.2), supposing that the function $p(t)$ is continuously differentiable. It is evident that for the monotone non-increasing function $p(t) = P\{\tau > t\}$, $p(0) \neq 0$, the condition $p(0) = 1$ follows from the relation (1.2).

After differentiating equation (1.2) with respect to the variable s and assuming $s = 0$, $p'(0) = -\lambda$, $t \geqslant 0$, we get the differential equation $p'(t) = -\lambda p(t)$, $t > 0$, the solution of which under the initial condition $p(0) = 1$, is (1.3), where the constant has to be positive, since $p(t) \leqslant 1$.

The distribution we found for the probability of the non-negative random variable τ is called *exponential distribution*. Its distribution function is given by (1.1) with the density

$$f(t) = \lambda e^{-\lambda t}, \quad t \geqslant 0.$$

The parameter $\lambda > 0$ has an obvious probabilistic illustration, namely

$$\frac{1}{\lambda} = M\tau = \int_0^\infty t f(t) dt$$

which is the mean value (the mathematical expectation) of the random variable τ.

We suppose that the existence of the so-called *half-life period* follows from the exponential formula (1.3) for the period of disintegration. The half-life period T is the period during which half of the initial material disintegrates. (It does not depend on the initial amount of **Ra**.)

Let us start with n **Ra** atoms. Each of them remains intact during the time t with probability $p(t)$, and the average number of **Ra** atoms remaining after time t is, according to (1.3),

$$n(t) = np(t) = ne^{-\lambda t}, \quad t \geqslant 0.$$

We specify that the number of **Ra** atoms remaining is a random variable $\nu(t)$, and we talk about the mathematical expectation

$n(t) = Mv(t)$.   It is obvious that the variable T, which is determined from the equation $n(T) = n/2$, does not depend on the initial number n of **Ra** atoms:

$$T = \ln 2/\lambda.$$

Problem: Let τ be a non-negative random variable with exponential distribution. We shall interpret τ as the "waiting time". Show that

(1.4) $P\{\tau > s + t \mid \tau > s\} = P\{\tau > t\},   s,t \geqslant 0,$

i.e. that after a time s, the "waiting time" has the same probability distribution as for the "waiting time" itself.

Problem: Let $\tau_1, ..., \tau_n$ be independent random variables with exponential distribution with corresponding parameters $\lambda_1, ..., \lambda_n$. Prove that the random variable $\tau = \min(\tau_1, ..., \tau_n)$ has exponential distribution with the parameter $\lambda = \lambda_1 + ... + \lambda_n$, that is

(1.5) $P\{\tau > t\} = e^{-(\lambda_1+...+\lambda_n)t},   t \geqslant 0.$

Prove that $\tau_1, ..., \tau_n$ are different from one another with probability 1, i.e. that the coincidence of some $\tau_1, ..., \tau_n$ has probability zero, and we can talk about a *first* (minimal) variable among $\tau_1, ..., \tau_n$.

Hint: Apply the equation $P\{\tau > t\} = P\{\tau_1 > t, ..., \tau_n > t\}$ and the condition that $\tau_1, ..., \tau_n$ are independent.

Problem: Let $\tau_1, ..., \tau_n$ be independent random variables, having exponential distribution with parameter λ, and $\tau = \min(\tau_1, ..., \tau_n)$. Let us denote by $\tau_1', ..., \tau_{n-1}'$ the variables different from 0 among $\tau_1 - \tau, ..., \tau_n - \tau$. Prove that $\tau_1', ..., \tau_{n-1}'$ are independent and that each variable $\tau' = \tau_k'$ has exponential probability distribution with the initial parameter λ:

(1.6) $P\{\tau' > t\} = e^{-\lambda t},   t \geqslant 0.$

Hint: Apply the invariance of the distribution of the variables $\tau_1', ...,$ τ_{n-1}' with regard to a rearrangement of $\tau_1, ..., \tau_n$ and the fact that with $\tau = \tau_n$,

$P\{\tau_1 - \tau_n > t_1, ..., \tau_{n-1} - \tau_n > t_{n-1} \mid \tau = \tau_n\}$

$= P\{\tau_1 > t_1 + \tau_n, ..., \tau_{n-1} > t_{n-1} + \tau_n \mid \tau_1 > \tau_n, ..., \tau_{n-1} > \tau_n\}$

$= e^{-\lambda t_1} ... e^{-\lambda t_{n-1}}.$

We return to our process of radioactive decay and we consider the number of α-particles $\xi(t)$ which are emitted during the period of time t. We consider the change of the variable $\xi(t)$ in the course of time t. If we have chosen the value $t_0 = 0$ at the start, we shall be dealing with the number of α-particles $\xi(t)$ which are emitted up to time t.

Let the number of **Ra** atoms at the start be equal to n and let τ_k^0 denote the time of disintegration of the kth **Ra** atom ($k = 1, ..., n$). We know that the random variables τ_k^0 have exponential probability distribution with the same parameter λ. Supposing that each **Ra** atom disintegrates independently of the state of the other atoms, we conclude that the time

$$\Delta_0 = \min(\tau_1^0, ..., \tau_n^0)$$

up to the appearance of the first α-particle obeys the exponential law with parameter $\lambda_0 = n\lambda$ [cf. (1.5)]. If we denote by $\xi(t)$ the state at time t of the process that we are considering, we can say that the initial state is $\xi(0) = 0$; the process remains in this state for the random time Δ_0, which is distributed exponentially with parameter $\lambda_0 = n\lambda$, but at time $\tau_0 = \Delta_0$, we observe the transition into the new state $\xi(\tau_0) = 1$. At time τ_0, there are still $n - 1$ **Ra** atoms. Denoting by τ_k' the time between the moment τ_0 and the decay of the kth of the remaining **Ra** atoms, we conclude that the process is in the state $\xi(\tau_0) = 1$ during the random time

$$\Delta_1 = \min(\tau_1', ..., \tau_{n-1}'),$$

which is distributed exponentially with parameter $\lambda_1 = (n-1)\lambda$ [we assume here (1.6)] and that, after a further time Δ_1, the transition into the new state $\xi(\tau_1) = 2$ occurs at time $\tau_1 = \tau_0 + \Delta_1$.

In general, the process that is in the state $\xi(\tau_i) = i + 1$ (independently of its behaviour up to time τ_i) stays in the state $i + 1$ for a random time Δ_{i+1}, which is distributed exponentially with parameter $\lambda_{i+1} = (n - i - 1)\lambda$, and then we have the transition into the new state $i + 2$, and so on.   The typical trajectory $x(t)$, $t \geqslant 0$, of the process is represented schematically in Figure 1.

Problem: Let $\Delta_0, \Delta_1, ...$ be random variables such that Δ_k does not depend on any $(\Delta_0, ..., \Delta_{k-1})$, $k = 1, 2, ...$. Show that $(\Delta_0, ..., \Delta_{k-1})$ and $(\Delta_k, ..., \Delta_n)$, $n > k$ are independent.

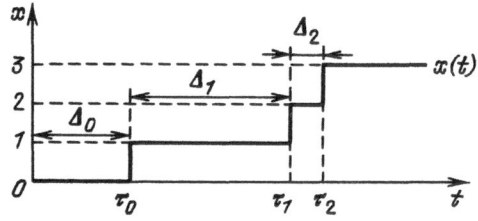

Figure 1

It is easy to conceive of the following generalization of the *random process* $\xi(t)$, $t \geqslant 0$, describing the evolution of some "system". Let us have a finite or denumerable number of possible states, denumerated by the number $i = 0,1, \dots$. At the beginning, the initial state is $\xi(0) = i_0$ and the process stays in this state for the random time Δ_0, distributed exponentially with parameter λ_{i_0}. Afterwards,

we observe at time $\tau_0 = \Delta_0$, independently of i_0 and Δ_0, the transition into some new state i_1 with corresponding probability $\pi_{i_0 i_1}$;

altogether, we have a chain of consecutive transitions

(1.7) $$\xi(0) = i_0 \rightarrow \xi(\tau_0) = i_1 \rightarrow \dots \rightarrow \xi(\tau_{k-1}) = i_k$$

into the states i_0, i_1, ..., i_{k-1}, in which the system remains for the corresponding periods Δ_0, Δ_1, ..., Δ_{k-1}; at time $\tau_{k-1} = \Delta_0 + \Delta_1 + \dots + \Delta_{k-1}$, the transition from the initial state i_{k-1} into the following state $\xi(\tau_{k-1}) = i_k$ occurs with corresponding probability $\pi_{i_{k-1} i_k}$,

independently of the past until this time; it stays in this state a random time Δ_k, distributed exponentially with parameter λ_{i_k},

whereupon the transition into a new state occurs, and so on.

Let us turn our attention to an important regularity. The whole behaviour of our process after time τ, where it is in some state $\xi(\tau) = i$, does not depend on the behaviour of this process before this time τ. In fact, the initial state being $i = \xi(\tau)$, the process remains, independently of the past, in state i during the random time Δ, distributed exponentially with parameter λ_i, and then passes with probability π_{ij} to the new state $j \neq i$ and so on. This property holds also for the behaviour of the process after any desired fixed time s, which is not necessarily a time at which a transition occurs, with known state $\xi(s) = i$ at the "running" time s: The behaviour of $\xi(t)$, $t \geqslant s$, in the "future" does not depend on the "past" $\xi(t)$, $t \leqslant s$.

To show this, we denote by $\tau \geqslant s$ the time in which the process leaves the initial state $\xi(s) = i$. As we know, the behaviour of the process after time τ with known state $\xi(\tau) = j$ does not depend on the past up to time τ, and we have only to show that the behaviour of the process in the interval $s \leqslant t \leqslant \tau$ does not depend on the past before time s.

The transition $i \rightarrow j$ from the initial state occurs at time τ (with probability π_{ij}), independently of the preceding situation, and hence we shall have proved the law that we are looking for, if we show that the time spent in the initial state $\xi(s) = i$ after the time s (i.e. the variable $\tau - s$) is independently of the "past" (before time s), distributed exponentially with parameter λ_i. Let $\tau' \leqslant s$ be the time at which we observe the transition of the process into the state $\xi(s) = i$ and $\Delta = \tau - \tau'$ the total time during which the process stays in this state. We know that, independently of the time τ', the variable Δ has exponential distribution with corresponding parameter λ ($\lambda = \lambda_i$ with $\xi(s) = i$ known).

If we consider the conditional probability of the *independent variables* τ' and Δ, we have

$$P\{\tau - s > t \mid \tau' \leqslant s, \tau > s\}$$

$$= P\{\Delta > t + (s - \tau') \mid \tau' \leqslant s, \Delta > s - \tau'\}$$

$$= \frac{P\{\Delta > t + (s - \tau'), \tau' \leqslant s\}}{P\{\Delta > s - \tau', \tau' \leqslant s\}} = e^{-\lambda t},$$

where, recall, $\lambda = \lambda_i$ with $\xi(s) = i$.

We shall now consider some examples of random proceses of the type described above.

Example: (*Poisson Process*). Once again we consider radioactive decay of **Ra**. It is known that this is a very slow process (experimental data give the value $T \approx 1600$ years for the constant of the half-life period), and considering the process of emission of α-particles during a relatively short compared to T time interval, we can assume that the amount of radium is constant. This simplifies the characteristics of our process $\xi(t)$, $t \geqslant 0$; obviously, the simplification affects the parameters λ_i, the values of which will now be

$$\lambda_i = n\lambda = \mu$$

for all $i = 0,1, \ldots$ (where n is the existing number of **Ra** atoms). It is easy to conceive of the generalization of this process $\xi(t)$ for the whole time axis $t \geqslant 0$. The whole behaviour is of the type that at the initial time $t = 0$ we have $\xi(0) = 0$, in which state the process remains for a random time Δ_0, distributed exponentially with corresponding parameter $\lambda_0 = \mu$, then, at time $\tau_0 = \Delta_0$, the transition into the state $\xi(\tau_0) = 1$ occurs. In this state, it spends the random time Δ_1, distributed exponentially with the same parameter μ, then, at time $\tau_1 = \tau_0 + \Delta_1$, we have the transition into the new state $\xi(\tau_1) = 2$. In general, if it has passed into the next state i at the random time $\tau_{i-1} = \Delta_0 + \ldots + \Delta_{i-1}$, the process remains, independently of the variables $\Delta_0, \ldots, \Delta_{i-1}$, in the state i for the random time Δ_i, distributed exponentially with parameter μ, and then we have, at time $\tau_i = \tau_{i-1} + \Delta_i$, the transition into the new state $i + 1$ and so on. A process of this type is called a *Poisson process* (with parameter μ).

Example: (*A single server system*) We imagine a service system that satisfies the demands that are made on it as follows: If the system is vacant, then, independently of the previous situation, the satisfaction of a demand requires a random time, distributed exponentially with parameter λ, and if the system is occupied, then the incoming demand is rejected and is no longer taken into consideration. We assume that the probability of more than one demand arriving simultaneously is equal to 0, and that, having satisfied the demand, the system waits, independently of the previous system, for the following demand during a random time, which has exponential distribution with parameter μ. Obviously, if we consider two states: $\xi(t) = 0$ for a vacant system, and $\xi(t) = 1$ for an occupied system at time t, we shall be dealing with a random process $\xi(t)$, $t \geqslant 0$ of the type (1.7) with parameters $\lambda_0 = \mu$, $\pi_{01} = 1$ and $\lambda_1 = \lambda$, $\pi_{10} = 1$.

Problem: Prove that a process of this type arises in the system described above if the stream of demands does not depend on the process of service and if it is a Poisson process (with parameter μ).

We return to the general process $\xi(t)$, $t \geqslant 0$ with parameters λ_i, π_{ij} which is described in (1.7). The attentive reader will have noticed that we considered this process only until the time

$$\tau = \Delta_0 + \Delta_1 + \dots = \lim_{n \to \infty} \tau_n \, ,$$

or, in other words, we were talking about the change of states $\xi(t)$ after a finite number of transitions.

Problem: Let $\tau = \Sigma_{k=0}^{\infty} \Delta_k$ be the sum of independent random variables Δ_k, which have exponential distribution with parameters λ_k, $k = 0,1,\dots$. Prove that $\tau = \infty$ with probability 1 if and only if

$$M\tau = \sum_{k=0}^{\infty} \frac{1}{\lambda_k} = \infty \, .$$

Hint: Apply the equality

$$Me^{-\tau} = \lim_{n \to \infty} \prod_{k=0}^{n} Me^{-\Delta_k} \, ,$$

where $Me^{-\tau} = 0$ if and only if $\tau = \infty$ with probability 1.

Problem: Let $\lambda_i \leqslant C$, $i = 0,1,\dots$. Prove that there is only a finite number of transitions (1.7) after a finite period of time with probability 1.

Hint: According to the supposition about independent events

$$\Delta_0 > h_0, \ i_0 \to i_1, \dots, i_{k-1} \to i_k, \ \Delta_{i_k} > h_k, \ i_k \to i_{k+1} \, ,$$

which describes the behaviour of the process after $k + 1$ consecutive transitions, the probability is

$$
\begin{aligned}
\text{(1.8)} \quad & P\{\Delta_k > h_0, \ i_0 \to i_1, \dots, i_{k-1} \to i_k, \ \Delta_k > h_k, i_k \to i_{k+1}\} \\
& = e^{-\lambda_{i_0} h_0} \cdot \pi_{i_0 i_1} \cdots \pi_{i_{k-1} i_k} \cdot e^{-\lambda_{i_k} h_k} \cdot \pi_{i_k i_{k+1}}
\end{aligned}
$$

with arbitrary $h_0, ..., h_k \geqslant 0$ and $i_0, i_1, ..., i_k, i_{k+1}$.

We assume that we have observed a finite number of transitions after a finite period of time with probability 1. Then, if the process is in any initial state $\xi(0) = i$, it is in some other state $\xi(t) = j$ reached by some chain of transitions at time $t > 0$ with the corresponding probability

$$p_{ij}(t) = \mathbf{P}\{\xi(t) = j \mid \xi(0) = i\}.$$

In fact, we have already seen that the behaviour of our process $\xi(t)$, $t \geqslant s$, with initial state $\xi(s) = i$, is the same as if it had started at time s, and then with arbitrary $t \geqslant s$

(1.9) $\mathbf{P}\{\xi(t) = j \mid \xi(s) = i\} = p_{ij}(t - s);$

This probability depends on the length of the interval (s,t), but does not depend on its location on the time axis.   (This shows the *homogeneity in time* of the process we are considering.) We repeat that with a given running state $\xi(s) = i$, the behaviour of the process $\xi(t)$, $t \geqslant s$ in the future does not depend on the course of the process $\xi(t)$, $t < s$ in the past, where the whole probabilistic behaviour in the future is completely determined by the initial state $\xi(s) = i$. Accordingly, the probability of existing in the state $\xi(t) = j$ at time $t > s$ with arbitrary $\xi(s_1) = i_1, ...,$ $\xi(s_m) = i_m$, $\xi(s) = i$ at the times $s_1 < ... < s_m < s$ does not depend on the conditions

$$\xi(s_1) = i_1, ..., \xi(s_m) = i_m$$

with given state $\xi(s) = i$, and we have

(1.10)
$$\mathbf{P}\{\xi(t) = j \mid \xi(s_1) = i_1, ..., \xi(s_m) = i_m, \xi(s) = i\}$$
$$= \mathbf{P}\{\xi(t) = j \mid \xi(s) = i\} = p_{ij}(t - s).$$

The property that we have expressed here for any i, j and $t \geqslant s$ is called the *Markov property*; $p_{ij}(t)$, $t \geqslant 0$ is called the *probability of transition* from the state i into the state j after time t, or, more simply, *the transition probability*.

Problem: Show that, given the Markov property (1.10), we have

(1.11)
$$\mathbf{P}\{\xi(t_1) = j_1, ..., \xi(t_n) = j_n \mid \xi(s_1) = i_1, ..., \xi(s_m) = i_m, \xi(s) = i\}$$
$$= \mathbf{P}\{\xi(t_1) = j_1, ..., \xi(t_n) = j_n \mid \xi(s) = i\}$$
$$= p_{ij_1}(t_1 - s) \cdots p_{j_{n-1}j_n}(t_n - t_{n-1})$$

for any desired states at arbitrarily connected times

$$s_1 < \cdots < s_m < s < t_1 < \cdots < t_n.$$

Hint: To obtain (1.11), we may apply the general formula (1.10), taking as corresponding running moments $t_{n-1}, ..., t_1, s$.

Further on, we show the method of characterization of the transition probabilities $p_{ij}(t)$ by parameters λ_i and that

$$(1.12) \qquad \lambda_{ij} = \lambda_i \pi_{ij}, \quad j \neq i.$$

Problem: Show that for small $h > 0$ the following asymptotic expressions hold:

$$(1.13) \qquad \begin{aligned} p_{ii}(h) &= 1 - \lambda_i h + o(h), \\ p_{ij}(h) &= \lambda_{ij} h + o(h), \quad j \neq i, \end{aligned}$$

where $o(h)/h \to 0$ for $h \to 0$ uniformly in i, j, and in the case of bounded parameters $\lambda_i \leqslant C$.

Hint: According to (1.8), with arbitrary $\xi(s) = i$ we have for the number of consecutive transitions after time h [we denote this number by $\nu(h)$]

$$P\{\nu(h) \geqslant 2 \mid \xi(s) = i\} \leqslant (1 - e^{-\lambda_i h}) \sum_{j \neq i} \pi_{ij}(1 - e^{-\lambda_j h}) = o(h).$$

Section 2
Homogeneous Markov Processes with a Countable Number of States
Kolmogorov's Differential Equations

We shall consider a system the state of which at time t is $\xi(t)$. Let the number of possible states be finite or countable. As usual, we design each state by a number $i = 0,1, \ldots$. We suppose that the process of the transition of the system from one state into another is caused by chance and obeys the laws described in (1.9), (1.10) with transition probabilities

(2.1)
$$p_{ij}(t) = \mathbf{P}\{\xi(t) = j \mid \xi(0) = i\}, \quad i,j = 0,1, \ldots$$
$$\left[\sum_i p_{ij}(t) = 1 \right].$$

We shall call $\xi(t)$, $t \geqslant 0$ a *homogeneous Markov process.*[1]

The model of a homogeneous Markov process $\xi(t)$, $t \geqslant 0$, with only a finite number of transitions from one state into another during a finite period of time was described in (1.7) from the point of view of the behaviour of a trajectory of the process (its development in time).

---

[1] For an arbitrarily chosen step width $h > 0$, the sequence of states

$$\xi_1 = \xi(h), \ldots, \xi_n = \xi(nh), \ldots$$

forms a so-called Markov chain with transition probability (for one step)

$$p_{ij} = p_{ij}(h), \quad i,j = 0,1, \ldots .$$

A Markov chain represents one of the most simplified and well-studied models of "discrete" probability theory. See, for example, W. Feller, An Introduction to Probability Theory and Its Applications, Vol. 1, Wiley & Sons, New York, 1968.

We shall now consider the general homogeneous Markov process $\xi(t)$, $t \geqslant 0$ with the transition probabilities (2.1). Suppose we have a probability distribution for the initial states:

(2.2)
$$P\{\xi(0) = i\} = p_i^0, \quad i = 0,1, \ldots$$
$$\left[\sum_i p_i^0 = 1 \right].$$

Then, according to the general formula (1.11), the joint probability distribution of the random variables $\xi(t_1)$, ..., $\xi(t_n)$ for any $0 = t_0 < t_1 \ldots < t_n$

will be:

(2.3)
$$P\{\xi(t_1) = j_1, \ldots, \xi(t_n) = j_n\}$$
$$= \sum_i p_i^0 p_{ij_1}(t_1 - t_0) \cdots p_{j_{n-1}j_n}(t_n - t_{n-1}).$$

In particular, the probability that the system is in the state j at time $t > 0$ is

(2.4) $p_j(t) = \sum_i p_i^0 p_{ij}(t), \quad j = 0,1, \ldots$.

Problem: Show that

(2.5) $p_j(s + t) = \sum_k p_k(s) p_{kj}(t), \quad s,t \geqslant 0.$

We examine the dependence of the transition probabilities $p_{ij}(t)$ on the time $t \geqslant 0$. It follows from the general formula (2.5), with $\xi(0) = i$ that for all $s,t \geqslant 0$

(2.6) $p_{ij}(s + t) = \sum_k p_{ik}(s) p_{kj}(t), \quad i,j = 0,1, \ldots$.

We introduce the matrix

$$P(t) = \{p_{ij}(t)\}, \quad t \geqslant 0.$$

Using this matrix, we can describe the system of equations (2.6) by the equation

(2.7) $P(s + t) = P(s) \cdot P(t), \quad s,t \geqslant 0.$

We suppose that the system which is in any state i stays in this state for some positive time τ, or, more precisely, $P\{\tau > 0\} = 1$. Then the transition probabilities $p_{ij}(t)$ are continuous for $t = 0$ where

(2.8) $p_{ij}(0) = \begin{cases} 1, & j = i, \\ 0, & j \neq i, \end{cases}$

or, in matrix form,

(2.9) $\lim_{h \to 0} P(h) = P(0) = I$,

where I is the unit matrix. In fact, for $h \to 0$,

$$p_{ii}(h) \geqslant P\{\tau > h\} \to P\{\tau > 0\} = 1,$$

$$p_{ij}(h) \leqslant 1 - p_{ii}(h) \to 0.$$

Let us prove the following theorem:

Theorem. *In the case of a finite number of states, the transition probabilities are continuously differentiable functions of t, and they satisfy the linear differential equations*

(2.10) $p_{ij}'(t) = \sum_k \lambda_{ik} p_{kj}(t),$

(2.11) $p_{ij}'(t) = \sum_k p_{ik}(t) \lambda_{kj},$

with constant coefficients

(2.12) $\lambda_{ij} = p_{ij}'(0), \quad i,j = 0,1, \dots .$

Proof. By condition (2.9), we have for the determinant det (h) of the matrix $P(h)$ that $\lim_{h \to 0} \det P(h) = 1$, and hence, the inverse matrix $P(h)^{-1}$ exists for sufficiently small $h > 0$, let us say for $h \leqslant \delta$. Looking at the condition (2.7) for $s,t \leqslant \delta$, we conclude that $P(h)^{-1}$ exists for $h = s + t \leqslant \delta$ (and so for all $h \geqslant 0$).

Moreover, with (2.7),

$$P(t + h) - P(t) = P(h)P(t) - P(t) = (P(h) - I)P(t) \to 0,$$

$$P(t - h) - P(t) = -P(h)^{-1}(P(h) - I)P(t) \to 0$$

for $h \to 0$, i.e. the matrix function $P(t)$ is continuous for all $t \geqslant 0$. The limit

$$\lim_{t_1,t_2 \to t} \frac{1}{t_2 - t_1} \int_{t_1}^{t_2} (s)ds = P(t)$$

exists with its determinant different from zero, det $P(t) \neq 0$, from which the existence of the inverse matrix

$$\left[\int_{t_1}^{t_2} P(s)ds \right]^{-1}$$

follows for a sufficiently small increment $t_2 - t_1 > 0$. Again, with equation (2.7) we have

$$(P(h) - I) \cdot \int_{t_1}^{t_2} P(s)ds = \int_{t_1}^{t_2} (P(h) - I) P(s)ds$$

$$= \int_{t_1+h}^{t_2+h} P(s)ds - \int_{t_1}^{t_2} P(s)ds = \int_{t_2}^{t_2+h} P(s)ds - \int_{t_1}^{t_1+h} P(s)ds$$

from which it follows that

$$\frac{P(h) - I}{h} = \left[\frac{1}{h} \int_{t_2}^{t_2+h} P(s)ds - \frac{1}{h} \int_{t_1}^{t_1+h} P(s)ds \right] \left[\int_{t_1}^{t_2} P(s)ds \right]^{-1}.$$

The existence of the limit

$$\lim_{h \to 0} \frac{P(h) - I}{h} = P'(0).$$

is obvious.

We assume

(2.13) $\Lambda = \{\lambda_{ij}\} = P'(0)$.

Applying equation (2.7) again, we get for $t > 0$ and for sufficiently small $h > 0$:

$$\frac{P(t+h) - P(t)}{h} = \frac{P(h) - I}{h} P(t) = P(t) \frac{P(h) - I}{h},$$

$$\frac{P(t-h) - P(t)}{-h} = P(h)^{-1} \frac{P(h) - I}{h} P(t)$$

$$= P(t) \frac{P(h) - I}{h} P(h)^{-1},$$

where $P(h)^{-1} \to I$ for $h \to 0$. It is obvious that the (continuous) derivative

(2.14) $P'(t) = \Lambda P(t) = P(t)\Lambda,$

exists, where the differential equations that we have here are nothing other than the corresponding system of differential equations (2.10) and (2.11) expressed in matrix form. The theorem is thus demonstrated.   □

It is well known that the solution of the differential equation (2.14) with the initial condition $P(0) = I$ is the matrix exponential function

(2.15) $P(t) = e^{\Lambda t}, \quad t \geqslant 0.$

Problem: Show that for the single server system (see the example on page 6) the transition probabilities are

$$p_{00}(t) = \left[1 - \frac{\lambda}{\lambda + \mu}\right] e^{-(\lambda + \mu)t} + \frac{\lambda}{\lambda + \mu}, \quad p_{01}(t) = 1 - p_{00}(t),$$

$$p_{11}(t) = \left[1 - \frac{\mu}{\lambda + \mu}\right] e^{-(\lambda + \mu)t} + \frac{\mu}{\lambda + \mu}, \quad p_{10}(t) = 1 - p_{11}(t).$$

The differential equations (2.10), (2.11) are justified under the most general conditions and for processes with an infinite number of states; (2.10) is called the *backward system* and (2.11) the *forward system of Kolmogorov's differential equations.*

We consider an arbitrary homogeneous Markov process with an infinite number of possible states $i = 0, 1, \ldots$. Let the transition probabilities $p_{ij}(t)$ be differentiable functions of $t \geqslant 0$ where the parameters of (2.10) satisfy the condition

$$(2.16) \qquad \sum_{j \neq i} \lambda_{ij} = -\lambda_{ii} = \lambda_i \,,$$

which corresponds to the equations

$$\sum_{j \neq i} p_{ij}(t) = 1 - p_{ii}(t), \quad t > 0.$$

[See also (1.12), (1.13).]

Theorem. *For differentiable[2] transition probabilities $p_{ij}(t)$ and under condition (2.16), the system of Kolmogorov's backward equations (2.10) is justified.*

Proof. According to (2.6), we have

$$\frac{p_{ij}(t+h) - p_{ij}(t)}{h} - \frac{p_{ii}(h) - 1}{h} p_{ij}(t) = \sum_{k \neq i} \frac{p_{ik}(h)}{h} p_{kj}(t).$$

Taking a finite number of terms, say n, which are all nonnegative, from the sum on the right-hand side and assuming $h \to 0$, then for $n \to \infty$, we get the inequality

$$p_{ij}'(t) - \lambda_{ii} p_{ij}(t) \geqslant \sum_{k \neq i} \lambda_{ik} p_{kj}(t).$$

For the left-hand side, with sufficiently small n and applying the approximation

---

[2]Assuming (1.13) (without the uniformity condition) we can easily see the differentiability of the $p_{ij}(t)$ after observing the continuity of the $p_{ij}(t)$. We argue similarly as in the proof given below by replacing *lim* by *lim sup* and *lim inf* respectively.

$$\lim_{h\to 0} \frac{1}{h} \sum_{k>n} p_{ik}(h)p_{kj}(t) \leqslant \lim_{h\to 0} \frac{1}{h} \sum_{k>n} p_{ik}(h) = \lim_{h\to 0} \frac{1}{h}\left[1 - \sum_{k\leqslant n} p_{ik}(h)\right]$$

$$= \lim_{h\to 0} \frac{1 - p_{ii}(h)}{h} - \lim_{h\to 0} \sum_{k\leqslant n, k\neq i} \frac{p_{ik}(h)}{h}$$

$$= \lambda_i - \sum_{k\leqslant n, k\neq i} \lambda_{ik},$$

we get the relation

$$p'_{ij}(t) - \lambda_{ii}p_{ij}(t) \leqslant \sum_{k\neq i} \lambda_{ik}p_{kj}(t) + \left[\lambda_i - \sum_{k\leqslant n, k\neq i} \lambda_{ik}\right],$$

where, according to the condition (2.16),

$$\left[\lambda_i - \sum_{k\leqslant n, k\neq i} \lambda_{ik}\right] \to 0 \quad \text{with} \quad n \to \infty.$$

Finally, we get the result we wanted to demonstrate,

$$p'_{ij}(t) - \lambda_{ii}p_{ij}(t) = \sum_{k\neq i} \lambda_{ik}p_{kj}(t). \qquad \square$$

Problem: Suppose the asymptotic expressions (1.13) hold for each j and uniformly for all states i from which transition into the state j is possible. Deduce the differential equations

$$(2.17) \qquad p'_j(t) = \sum_k p_k(t)\lambda_{kj}, \quad j = 0,1, \dots$$

for the probabilities $p_j(t) = P\{\xi(t) = j\}$, $t \geqslant 0$.

Hint: Apply the equation

$$\frac{p_j(t+h) - p_j(t)}{h} = p_j(t)\frac{p_{jj}(h) - 1}{h} + \sum_{k\neq j} p_k(t)\frac{p_{kj}(h)}{h},$$

where the series $\sum_k p_k(t)$ is convergent and

$$\frac{p_{kj}(h)}{h} = \lambda_{kj} + o(1)$$

uniformly for all k.

We assume that, with the condition $\xi(0) = i$, (2.17) gives us the forward system of differential equations (2.11).

Example: (*Poisson process*) We described the same process when we considered the stream of α-particles. It is obvious from this description that it is a homogeneous Markov process, satisfying the conditions (1.13), and hence the system of differential equations (2.17) is justified with the parameters

$$\lambda_{ij} = \begin{cases} -\lambda, & j = i, \\ \lambda, & j = i + 1, \\ 0, & j \neq i, \ i + 1. \end{cases}$$

In our example, the system (2.17) looks like

$$p_0'(t) = -\lambda p_0(t),$$

$$p_k'(t) = \lambda p_{k-1}(t) - \lambda p_k(t), \quad k = 1, 2, \dots .$$

Setting

$$f_k(t) = e^{\lambda t} p_k(t), \quad k = 0, 1, \dots ,$$

we get

$$f_0'(t) = \lambda f_0(t) + e^{\lambda t} p_0'(t) = \lambda f_0(t) - \lambda f_0(t) = 0,$$

$$f_k'(t) = \lambda f_k(t) + e^{\lambda t} p_k'(t) = \lambda f_k(t) + \lambda e^{\lambda t} p_{k-1}(t) - \lambda e^{\lambda t} p_k'(t)$$

$$= \lambda f_{k-1}(t), \quad k = 1, 2, \dots ,$$

and hence

$$f_0(t) = 1, \quad f_1(t) = \lambda t, \dots, f_k(t) = \frac{(\lambda t)^k}{k!}, \dots$$

subject to the initial conditions

$$f_0(0) = p_0(0) = 1, \quad f_k(0) = p_k(0) = 0, \quad k = 1, 2, \dots .$$

Our result is the well-known *Poisson distribution*

$$p_k(t) = \frac{(\lambda t)^k}{k!} e^{-\lambda t}, \quad k = 0, 1, \dots .$$

Problem: Show that the Poisson process is homogeneous with regard to the states of the process in the sense that the transition from state i into a state $j > i$ during the time t has the same probability as the transition from 0 into the state $k = j - i$, i.e.

$$p_{ij}(t) = p_k(t) = \frac{(\lambda t)^k}{k!} e^{-\lambda t}, \quad k = j - i.$$

Hint: Apply the forward system (2.11).

Problem: Show that for any $0 < t_1 \dots < t_n$, the increments $\xi(t_{k+1}) - \xi(t_k)$, $k = 0, \dots, n-1$ of the Poisson process are independent random variables, where the increment on the interval (s, t) has Poisson distribution with parameter $\mu = \lambda(t - s)$:

$$P\{\xi(t) - \xi(s) = k\} = \frac{(\lambda(t-s))^k}{k!} \, e^{-\lambda(t-s)} \, , \quad k = 0,1, \dots .$$

Hint: Apply the equation

$$P\{\xi(t_1) - \xi(0) = i_1, \dots, \xi(t_n) - \xi(t_{n-1}) = i_n\}$$

$$= P\{\xi(t_1) = i_1, \dots, \xi(t_{n-1}) = i_1 + \dots + i_{n-1}, \xi(t_n) = i_1 + \dots + i_n\}$$

and the general formula (2.3) with $\xi(0) = 0$, $p_0^0 = 1$.

Section 3
Homogeneous Markov Processes with a Countable Number of States
Convergence to a Stationary Distribution

Considering a homogeneous Markov process $\xi(t)$, $t \geqslant 0$, with a countable number of states j, we say that the probability distribution $\{p_j^*\}$ is *stationary*, if

(3.1) $\qquad p_j^* = \sum_k p_k^* p_{kj}(t), \quad j = 0,1, \ldots ,$

where $\{p_{kj}(t)\}$ are the transition probabilities of the process. It follows from the general formulas (2.3) and (2.4) with a stationary initial distribution $p_i^0 = p_i^*$ that

(3.2) $\qquad \mathbf{P}\{\xi(t_1 + t) = i_1, \ldots, \xi(t_n + t) = i_n\} = \mathbf{P}\{\xi(t_1) = i_1, \ldots, \xi(t_n) = i_n\},$

i.e. the probability distribution of arbitrary random variables $\xi(t_1), \ldots, \xi(t_n)$ does not change under a shift to time $t \geqslant 0$. In particular, the probability distribution of the variables $\xi(t)$ will be exactly the same for all t:

(3.3) $\qquad p_j(t) = \mathbf{P}\{\xi(t) = j\} = p_j^* , \quad j = 0,1, \ldots .$

We suppose that for the process $\xi(t)$, $t \geqslant 0$, there exists at least one state j_0 into which a transition is possible from an arbitrary state i after a time $h > 0$ with corresponding probabilities

(3.4) $\qquad p_{ij_0}(h) \geqslant \delta > 0, \quad i = 0,1, \ldots$

(we emphasize that in this condition $h > 0$ is always the same for all states i). Then the following theorem is justified:

Theorem. *There exists a unique stationary distribution* $\{p_j^*\}$, *and for* $t \to \infty$ *we have*

$$p_j(t) = P\{\xi(t) = j\} \to p_j^* , \quad j = 0,1, \dots .$$

Moreover,

(3.5) $\quad |p_j(t) - p_j^*| \leqslant (1 - \delta)^{(t/h)-1}$

uniformly for all states, independently of the initial probability distribution.

Proof: Denote by Π^0 the set of all probability distributions $p^0 = \{p_i^0\}$. The transformation (2.4) with the matrix $P(t)$ changes $p^0 = \{p_j^0\}$ into $p(t) = \{p_j(t)\}$; we denote by $\Pi(t) = \Pi^0 P(t)$ the set of all distributions

$$p(t) = p^0 P(t), \quad p^0 \in \Pi^0.$$

From equation (2.7) we can see that

$$\Pi(s + t) = \Pi(s) P(t) \subseteq \Pi^0 P(t) = \Pi(t)$$

for all $s,t \geqslant 0$, and so the sets $\Pi(t)$, $t \geqslant 0$ turn out to be contained in one another. We denote by Π^* their intersection:

$$\Pi^* = \lim_{t \to \infty} \Pi(t) \quad \left(= \bigcap_{t \geqslant 0} \Pi(t)\right).$$

It is obvious that the limit set Π^* is invariant with respect to a transformation by $P(t)$, $t \geqslant 0$ since

$$\Pi^* P(t) \subseteq \Pi(s) P(t) = \Pi(s + t), \quad s \geqslant 0,$$

and

$$\Pi^* P(t) \subseteq \bigcap_{s \geqslant 0} \Pi(s + t) = \Pi^*.$$

We observe that each stationary distribution $p^* = \{p_j^*\}$ is a point of the set Π^*, since the stationarity of the distribution p^* shows nothing other than its invariance with respect to a transformation by $P(t)$:

$$p^* P(t) = p^*, \quad t \geqslant 0 \quad \text{[see (3.1)]}.$$

We introduce a distance between the "points" p', $p'' \in \Pi^0$ which we define by

$$\|p' - p''\| = \sup_i |p_i' - p_i''|,$$

and we consider the "diameter"

$$\text{diam } \Pi(t) = \sup_{p',p'' \in \Pi(t)} \|p' - p''\|$$

of the sets $\Pi(t)$ contained in one another. We will show that under condition (3.4) the transformation (2.4) is "contracting" and

$$\lim_{t\to\infty} \text{diam } \Pi(t) = 0.$$

If we have this result, then the limit set Π^* (if it is nonempty!) consists of one unique point $p^* = \{p_i^*\}$. We know that the set Π^* is invariant, and in the case that Π^* consists only of one unique point p^*, this shows the invariance (stationarity) of $p^* \doteq \{p_i^*\}$.

(In the case of a *finite* number of states and if we consider the distributions $p = \{p_i\}$ as points in a vector space with corresponding coordinates $p_i \geqslant 0$, $\Sigma_i p_i = 1$, we are dealing with *compact sets* $\Pi(t)$, $t \geqslant 0$ which are contained in one another and hence their intersection Π^* is nonempty! In the general case of an *infinite* number of states, supplementary proofs are necessary to deduce that the set Π^* is nonempty, and we demonstrate our theorem in another way.)

Let us look at the proof of our estimation (3.5). We set

$$r_j(t) = \inf_i p_{ij}(t), \quad R_j(t) = \sup_i p_{ij}(t).$$

where $r_j(t)$ and $R_j(t)$ give us a lower and an upper bound, respectively, for the probabilities

$$p_j(t) = \sum_i p_i^0 p_{ij}(t) \begin{cases} \geqslant \sum_i p_i^0 r_j(t) = r_j(t), \\[2mm] \leqslant \sum_i p_i^0 R_j(t) = R_j(t). \end{cases}$$

We observe that the diameter of the set $\Pi(t)$ may be expressed as:

$$\begin{aligned}
\text{diam } \Pi(t) &= \sup_{p',p''\in\Pi^0} \| p' \, P(t) - p'' \, P(t)\| \\
&= \sup_i \sup_{p',p''\in\Pi^0} \left\| \sum_i p_i' p_{ij}(t) - \sum_k p_k'' p_{kj}(t) \right\| \\
&= \sup_j (R_j(t) - r_j(t)).
\end{aligned}$$

We can show that the lower bound $r_j(t)$ increases monotonically, but also that the upper bound $R_j(t)$ decreases monotonically. In fact, for arbitrary $t \geqslant s$, we have

$$r_j(t) = \inf_i \left[\sum_k p_{ik}(t-s) p_{kj}(s) \right]$$

$$\geqslant \inf_i \left[\sum_k p_{ik}(t-s) r_j(s) \right] = r_j(s),$$

$$R_j(t) = \sup_i \left[\sum_k p_{ik}(t-s) p_{kj}(s) \right]$$

$$\leqslant \sup_i \left[\sum_k p_{ik}(t-s) R_j(s) \right] = R_j(s).$$

Furthermore,

$$R_j(t) - r_j(t) = \sup_{\alpha,\beta} [p_{\alpha j}(t) - p_{\beta j}(t)]$$

$$= \sup_{\alpha,\beta} \sum_k [p_{\alpha k}(h) - p_{\beta k}(h)]p_{kj}(t - h), \quad t \geqslant h.$$

Here $\sum_k p_{\alpha k}(h) = \sum_k p_{\beta k}(h) = 1$, and we can take out of the sum

$$0 = \sum_k [p_{\alpha k}(h) - p_{\beta k}(h)],$$

which is equal to 0, the sums which represent the positive and the negative terms of the sum:

$$\sum_k^+ [p_{\alpha k}(h) - p_{\beta k}(h)] = -\sum_k^- [p_{\alpha k}(h) - p_{\beta k}(h)].$$

It is easy to understand that with condition (3.4)

$$\sum_k^+ [p_{\alpha k}(h) - p_{\beta k}(h)] = \frac{1}{2} \sum_k |p_{\alpha k}(h) - p_{\beta k}(h)| \leqslant \frac{1}{2}(2-2\delta) = 1 - \delta$$

and, hence,

$$R_j(t) - r_j(t) \leqslant \sup_{\alpha,\beta} \left\{ \sum_k^+ [p_{\alpha k}(h) - p_{\beta k}(h)]R_j(t - h) \right.$$

$$\left. + \sum_k^- [p_{\alpha k}(h) - p_{\beta k}(h)]r_j(t - h) \right\}$$

$$= \sup_{\alpha,\beta} \sum_k^+ [p_{\alpha k}(h) - p_{\beta k}(h)](R_j(t-h) - r_j(t-h))$$

$$\leqslant (1 - \delta)(R_j(t - h) - r_j(t - h)).$$

From this, we have

$$R_j(t) - r_j(t) \leqslant (1 - \delta)^n (R_j(t - nh) - r_j(t - nh)) \leqslant (1-\delta)^{(t/h)-1}.$$

where n denotes the fraction t/h.

It is obvious that the lower and the upper bounds of the probabilities $p_j(t)$,

$$r_j(t) \leqslant p_j(t) \leqslant R_j(t),$$

approach each other for $t \to \infty$ (uniformly for all j) and that the unique limit

$$p_j^* = \lim_{t\to\infty} r_j(t) = \lim_{t\to\infty} p_j(t) = \lim_{t\to\infty} R_j(t).$$

exists.

We remember that the lower bound $r_j(t)$ increases monotonically and that the upper bound $R_j(t)$ decreases monotonically with increasing t so that the limit value p_j^* for all $t \geqslant 0$, lies between exactly the same bounds $r_j(t) \leqslant p_j(t)^* \leqslant R_j(t)$, and the same holds for the probabilities $p_j(t)$: therefore

$$|p_j(t) - p_j^*| \leqslant R_j(t) - r_j(t) \leqslant (1 - \delta)^{(t/h)-1} \,,$$

and so the estimation (3.5) is justified. To complete the proof of the theorem, we have to demonstrate that the limit values $\{p_j^*\}$ do, in fact, describe the stationary probability distribution. (We note that we do not yet know whether $p^* = \{p_j^*\}$ belongs to the limit set Π^*.)

It is obvious that $\Sigma_j p_j \leqslant 1$, and, hence, in this sum, for an arbitrary *finite* number of terms, we have

$$\sum_j p_j^* = \lim_{t \to \infty} \sum_j p_j(t) \leqslant 1.$$

Here $\Sigma_j p_j \neq 0$ because we have for the lower bound $r_{j_0}(h)$, according to condition (3.4)

$$p_{j_0}^* \geqslant r_{j_0}(h) \geqslant \delta.$$

Moreover, it follows from the general formula (2.6) for $s \to \infty$ that

$$p_j^* \geqslant \sum_k p_k^* p_{kj}(t), \quad t \geqslant 0.$$

In fact, the equality must hold because if the strong inequality hold for only one j, then we would have

$$\sum_j p_j^* > \sum_j \sum_k p_k^* p_{kj}(t) = \sum_k p_k^* \sum_j p_{kj}(t) = \sum_k p_k^*.$$

Taking the probability distribution

$$p_j^0 = p_j^* \mid \sum_k p_k^*, \quad j = 0,1, ...,$$

we convince ourselves that it is stationary:

$$p_j^0 = \sum_k p_k^0 p_{kj}(t), \quad t \geqslant 0.$$

With this as the initial distribution, we get $p_j(t) = p_j^0$, and we conclude that

$$p_j^* = \lim_{t \to \infty} p_j(t) = p_j^0, \quad j = 0,1, ... \,.$$

The relations that we have proved here are justified for the arbitrary stationary distribution $\{p_j^0\}$, and so there exists a unique stationary distribution $p^0 = p^*$.

We point out that the uniqueness was proved earlier because diam $\Pi(t) \to 0$ for $t \to \infty$ where, according to our estimation,

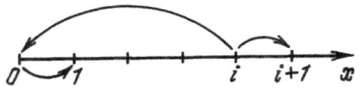

Figure 2

$$\text{diam } \Pi(t) \leqslant \sup_j (R_j(t) - r_j(t)) \leqslant (1 - \delta)^{(t/h)-1}.$$

The proof is complete. □

Let us now suppose that for the transition probabilities the forward system of differential equations (2.17) is justified. Taking the stationary distributions $\{p_j^*\}$, we get for the constants $p_j(t) = p_j^*$ the system of linear equations

(3.6) $\sum_k p_k^* \lambda_{kj} = 0, \quad j = 0,1, \dots .$

Problem: Let us consider a process of the type that we described in (1.7) and that is schematically sketched in Fig. 2: from the state i ($i = 1,2, \dots$) the system directly changes either to the next state $i + 1$ or to the state 0 from where it directly changes to the state 1. According to formula (1.12),

$$\lambda_{ij} = 0, \quad j \neq 0, i, i + 1.$$

Under which conditions for the parameters $\lambda_{i,0}$ and $\lambda_{i,i+1}$ ($\lambda_{i,0} + \lambda_{i,i+1} = -\lambda_{ii} = \lambda_i$) does the stationary distribution exist and what is it like?

Hint: Apply the system of equations (3.6).

Problem: Let us have only a finite number of possible states of a homogeneous Markov process which, with an arbitrary initial state i, can be in any other state j (after some time $t > 0$) with some positive probability $p_{ij}(t) > 0$ which is proper for each pair (i,j). Show that the condition (3.4) holds.

Hint: Apply the relation (2.6) and deduce that $p_{ij}(s + t) > 0$ for all $t \geqslant 0$ if $p_{ij}(s) > 0$.

Example: *(multi-server system)* Imagine a service system that is analogous to the system which was described on page 6, but that has, instead of only one line of service, several, say n, lines of service. Each of these n service-lines satisfies the demand that comes in on this line during a certain random time, which is exponentially distributed with parameter λ. Under the condition that j lines are occupied, the waiting time until one of them is free is

$$\tau = \min(\tau_1, \ldots, \tau_j),$$

where $\tau_1, \ldots, \tau_j$ are independent random variables, which denote the waiting time until the end of the service on each of the j occupied lines and which all have the same exponential distribution with parameter λ. As we know, the variable τ is distributed exponentially with parameter $j\lambda$. Accordingly, the change of the number $\xi(t)$ of lines that are occupied at time t, in the course of time, is a homogeneous Markov process with $n + 1$ states $j = 0,1, \ldots, n$ for which the parameters (2.12) may be

$$\lambda_{0j} = \begin{cases} -\mu, & j = 0, \\ \mu, & j = 1, \\ 0, & j \neq 0,1 \end{cases} \qquad \lambda_{nj} = \begin{cases} n\lambda, & j = n - 1, \\ -n\lambda, & j = n, \\ 0, & j \neq n - 1, n, \end{cases}$$

$$\lambda_{ij} = \begin{cases} i\lambda, & j = i - 1, \\ -i\lambda - \mu, & j = i, \\ \mu, & j = i + 1, \\ 0, & j \neq i - 1, i, i + 1, \end{cases} \qquad 0 < i < n,$$

where μ is, we recall, the parameter of the exponential distribution of the waiting time of the following demand. The system of equations (3.6) gives us

$$-\mu p_0^* + \lambda p_1^* = 0,$$

$$\mu p_{i-1}^* - (\mu + i\lambda)p_i^* + (i + 1)\lambda p_{i+1}^* = 0, \quad 0 < i < n,$$

$$\mu p_{n-1}^* - n\lambda p_n^* = 0.$$

We easily see that

$$p_j^* = \frac{\dfrac{1}{j!}\left(\dfrac{\mu}{\lambda}\right)^j}{\displaystyle\sum_{k=0}^{n} \dfrac{1}{k!}\left(\dfrac{\mu}{\lambda}\right)^k}, \quad j = 0, \ldots, n.$$

This expression is well known as *Erlang's formula*. It is clear that the system may accomplish the transition from an arbitrary state into any other state and that condition (3.4) holds. Therefore, the probability distribution of $\xi(t)$ converges with $t \to \infty$ to the stationary distribution we found above [see the estimation (3.5)].

Section 4
Branching Processes
Method of Generating Functions

We now consider a *branching process* $\xi(t)$, $t \geqslant 0$, for example the transformation of one type of particles which follows the principle that each particle existing at time s is, independently of the past (up to time s), transformed into n particles with probability $p_n(t)$, $n = 0,1, \ldots$. We will characterize the state of the process at time t by the total number $\xi(t)$ of particles existing at this moment (we do not exclude the possibility $n = \infty$).

Accordingly, assuming the condition $\xi(s) = k$, the number of particles after time t will be

(4.1) $\xi(s + t) = \xi_1(t) + \ldots + \xi_k(t),$

where $\xi_i(t)$ denotes the number of particles generated by the transformation of the ith initial particle after time t. The independent random variables $\xi_1(t)$, ..., $\xi_k(t)$ have the same probability distribution:

$$\mathbf{P}\{\xi_i(t) = n\} = p_n(t), \quad n = 0,1, \ldots, \infty .$$

We consider $\xi(t)$, $t \geqslant 0$ as a homogeneous Markov process with transition probabilities

$$p_{kn}(t) = \mathbf{P}\{\xi_1(t) + \ldots + \xi_k(t) = n\}, \quad k \neq 0, \infty ,$$

and for $k = 0, \infty$ we take

$$p_{00}(t) = 1, \quad p_{0n}(t) = 0, \quad n \neq 0,$$

$$p_{\infty\infty}(t) = 1, \quad p_{\infty n}(t) = 0, \quad n \neq \infty .$$

Assume that $p_{1n}(t) = p_n(t)$, $n = 0,1, \ldots, \infty$. Differentiating with respect to $t \geqslant 0$ and introducing $\lambda_{1n} = p'_{1n}(0)$, it follows, applying condition (2.16), that

(4.2) $\sum_{n \neq 1} \lambda_{1n} = -\lambda_{11} = \lambda_1.$

If we consider the transition process from one state to another as we did in Section 1 [see (1.7)], we can say that the constants λ_{1k} characterize the probabilities of the direct transition from state 1 into another new state $k \neq 1$. In particular, the transition $1 \rightarrow k$ is impossible, with $\lambda_{1k} = 0$. We will assume that $\lambda_{1\infty} = 0$. We know that condition (4.2) justifies the differential equations

(4.3) $p'_{1n}(t) = \sum_k \lambda_{1k} p_{kn}(t), \quad n = 0,1, \dots .$

For these equations Kolmogorov's backward system of differential equations holds.

We introduce the *generating function* of a variable z, $0 \leqslant z < 1$, which is determined by the series

$$F_k(t,z) = \sum_n p_{kn}(t) z^n ,$$

where the sum is over $n = 0,1, \dots$. The equations (4.3) and (4.2) imply $|p'_{1n}(t)| \leqslant 2\lambda_1$, and for every fixed z, $0 \leqslant z \leqslant 1$, we have

$$\sum_n p'_{1n}(t) z^n = \sum_n p'_{1n}(t) z^n = \sum_k \lambda_{1k} \sum_n p_{kn}(t) z^n .$$

This justifies the following differential equation corresponding to the generating function $F_1(t,z)$:

$$\frac{d}{dt} F_1(t,z) = \sum_k \lambda_{1k} F_k(t,z).$$

The functions $F_k(t,z)$, $k \neq 0$ (assuming $z^\infty = 0$, $0 \leqslant z < 1$) represent the mathematical expectation:

$$F_k(t,z) = M z^{\xi_1(t)+\dots+\xi_k(t)} ,$$

where the random variables $\xi_1(t), \dots, \xi_k(t)$ from (4.1) are independent and have the same distribution:

$$M z^{\xi_1(t)+\dots+\xi_k(t)} = M z^{\xi_1(t)} \dots M z^{\xi_k(t)} .$$

Hence

(4.4) $F_k(t,z) = F_1(t,z)^k, \quad k = 1,2, \dots .$

As $F_0(t,z) = 1$, the differential equations for the generating function $F(t,z) = F_1(t,z)$ can be written in the form

(4.5) $\frac{d}{dt} F(t,z) = \sum_k \lambda_{ik} F(t,z)^k.$

We introduce the function

(4.6) $\qquad f(x) = \sum_{k} \lambda_{ik} x^k, \quad 0 \leqslant x \leqslant 1.$

When equation (4.5) holds, the generating function $F(t,z)$ for fixed z, $0 \leqslant z < 1$, is a solution of the differential equation

(4.7) $\qquad \dfrac{dx}{dt} = f(x), \quad t \geqslant 0.$

Because of the fact that $F(0,z) = z$, the generating function $F(t,z)$ coincides for every z, $0 \leqslant z < 1$, with the solution $x = x(t)$ of the equation (4.7), the initial condition being $x(0) = z$.

Instead of equation (4.7) it is convenient to consider the equivalent equation for the inverse function of $x = x(t)$, that is for $t = t(x)$:

$$\dfrac{dt}{dx} = \dfrac{1}{f(x)}, \quad 0 \leqslant x \leqslant 1,$$

writing the solution of this equation in the form

(4.8) $\qquad t(x) = \displaystyle\int_{z}^{x} \dfrac{du}{f(u)}, \quad 0 \leqslant x \leqslant 1.$

Example: Let the transition densities be equal to

$$\lambda_{10} = \lambda, \quad \lambda_{11} = -\lambda, \quad \lambda_{1k} = 0 \text{ for } k > 1.$$

Then $f(x) = \lambda(1 - x)$ and

$$t(x) = \int_{z}^{x} \dfrac{du}{f(u)} = -\dfrac{1}{\lambda}[\ln(1 - x) - \ln(1 - z)].$$

We can easily find the function $F(t,z)$ from this relation. In fact,

$$\ln(1 - F) = -\lambda t + \ln(1 - z)$$

and

$$F(t,z) = 1 - e^{-\lambda t}(1 - z).$$

The probabilities $p_n(t) = p_{1n}(t)$ which are determined by the expansion

$$F(t,z) = \sum_{n} p_n(t) z^n$$

are equal to

$$p_0(t) = 1 - e^{-\lambda t},$$

$$p_1(t) = e^{-\lambda t},$$

$$p_n(t) = 0, \quad \text{for } n > 1.$$

Problem: Suppose that the particles multiply by "division in halves" and that

$$\lambda_{10} = 0, \quad \lambda_{11} = -\lambda, \quad \lambda_{12} = \lambda, \quad \lambda_{1k} = 0, \quad \text{for } k > 2.$$

Find the generating function $F(t,z)$ and the corresponding probabilities

$$p_n(t) = p_{1n}(t), \quad n = 0,1, \dots .$$

We continue to consider the differential equations (4.7), (4.8), where the function $f(x)$ is determined by formula (4.6). We can see from this formula that

$$f''(x) = \sum_{k \geqslant 2} k(k - 1)\lambda_{1k}x^{k-2} \geqslant 0 \quad \text{for } 0 \leqslant x < 1,$$

so that the function $f(x)$ is convex and its derivative $f'(x)$ increases monotonically on the interval $0 < x < 1$. The value $x = 1$ is a root of the equation $f(x) = 0$, because $\sum_{k=0}^{\infty}\lambda_{1k} = 0$.

There may exist another root $x = \alpha$ of this equation, and hence the graph of the function $f(x)$ looks like Figure 3. Suppose that there is a root $x = \alpha, \; 0 < \alpha < 1$. It defines the integral curve $x(t) \equiv \alpha$ of the differential equations (4.7), (4.8). We choose the integral curve running through the point $t = 0, \; x = z \; (0 \leqslant z < \alpha)$. Since the derivative $f'(\alpha)$ is finite and since for $x \sim \alpha$ the function $f(x)$ is approximately equal to $f'(\alpha)(x - \alpha)$, it follows that along the integral curve the value of

$$t(x) = \int_z^x \frac{du}{f(u)}$$

increases unboundedly for $x \to \alpha.$

This curve does not intersect the other integral curve $x(t) \equiv \alpha$ anywhere. The function $f(x)$ is positive in the interval $0 \leqslant x < \alpha$ and hence $x(t)$ is monotone increasing for $t \to \infty$ along the integral curve, but remains bounded by $x = \alpha.$ As a bounded monotonic function, $x(t)$ has a limit $\beta = \lim_{t\to\infty}x(t), \; z \leqslant \beta < \alpha.$ But for $x \to \beta$, the continuous function $f(x)$ has the limit $f(\beta)$:

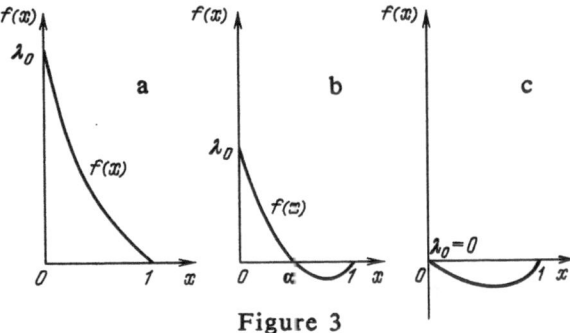

Figure 3

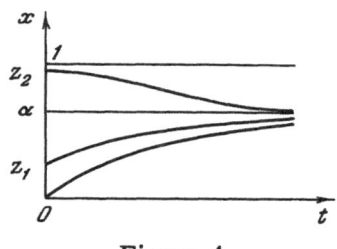

Figure 4

$$f(\beta) = \lim_{t\to\infty} f(x(t)) = \lim_{t\to\infty} x'(t).$$

It is clear that the value of $f(\beta)$ must be equal to zero, because otherwise the function

$$x(t) = z + \int_0^t f(x(s))ds$$

will increase unboundedly for $t \to \infty$. Hence β is a root of the equation $f(x) = 0$ and coincides with α : $\beta = \alpha$. Consequently, all integral curves $x = x(t)$ which run for $t = 0$ through the point $x = z$, $0 \leqslant z < \alpha$, are monotone increasing and have for $t \to \infty$ the limit

(4.9) $\lim_{t\to\infty} x(t) = \alpha$.

Integral curves which for $t = 0$ pass through the point $x = z$, $\alpha < z < 1$ ($0 \leqslant \alpha < 1$) behave in an analogous way. The only difference is the fact that $x(t)$ is monotone decreasing, since the derivative $x'(t) = f(x(t))$ is negative ($f(x) \leqslant 0$ for $\alpha \leqslant x < 1$). The whole graph of the integral curves for values of z in the interval $0 \leqslant z < 1$ is shown in Figure 4. Obviously, this picture is more simple for $\alpha = 0$.

It is necessary to consider the case $z = 1$ in particular. The corresponding integral curve is of the form $x(t) \equiv 1$. (Recall that $f(1) = 0$.) Let $f(x)$ take the value 0 for $x = 1$, so that the function $1/f(x)$ is not integrable in the neighbourhood of the point $x = 1$. That is, we have $\alpha < 1$ and

(4.10) $\displaystyle\int_{x_0}^1 \frac{du}{f(u)} = -\infty, \quad \alpha < x_0 < 1.$

We take an arbitrary integral curve; suppose that we have for $x = x_0$ the value $t_0 = t(x_0) \geqslant 0$ and that the corresponding curve is

$$t(x) = t_0 + \int_{x_0}^x \frac{du}{f(u)}.$$

It is obvious that our curve, which lies in the domain $t \geqslant 0$, does not intersect the integral curve $x = 1$, since for $x = 1$ we would have

$$t(1) = t_0 + \int_{x_0}^1 \frac{du}{f(u)} = -\infty ;$$

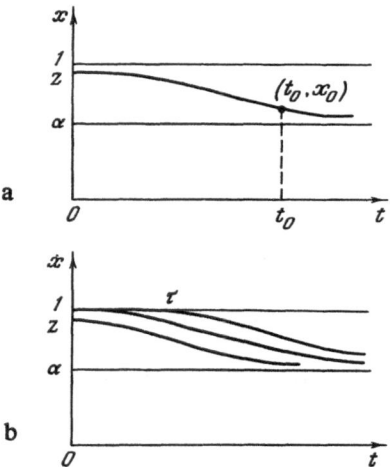

Figure 5

In particular, for $t = 0$ we have the value $x(0) = z < 1$. Therefore, $x(t) \equiv 1$ is the unique integral curve which passes through the point $t = 0$, $x = 1$. Suppose now that for the function $1/f(x)$ for $x = 1$ we have

$$(4.11) \qquad \int_{x_0}^{1} \frac{dx}{f(x)} > -\infty .$$

Then the integral curve

$$t(x) = t_0 + \int_{x_0}^{x} \frac{du}{f(u)}$$

intersects, for sufficiently large $t_\theta > 0$, the integral curve $x(t) \equiv 1$, which is its tangent at a certain point $t = \tau$, $x = 1$, where

$$\tau = t_0 + \int_{x_0}^{1} \frac{du}{f(u)} \geqslant 0$$

(Figure 5). In this case, an entire family of integral curves $x_\tau(t)$ passes through the point $t = 0$, $x = 1$, each curve corresponding to a value $\tau \geqslant 0$. Among them is the integral curve $x_0(t)$, which corresponds to the value $\tau = 0$, and which has the property that the curve $x_0(t)$ lies below all other integral curves $x_\tau(t)$:

$$x_0(t) < x_\tau(t), \quad 0 < t < \infty .$$

This can be explained by the fact that, within the domain $0 \leqslant x < 1$, $0 \leqslant t < \infty$, the solution of the corresponding differential equation is unique, and that integral curves do not intersect each other in this domain. It is easy to see that the integral curve $x_0(t)$ is the limit curve of the other integral curves $x(t,z)$ lying below it and passing through the corresponding point $t = 0$, $x = z$, where $0 \leqslant z < 1$:

(4.12) $x_0(t) = \lim\limits_{z \to 1} x(t,z).$

The study of the differential equations (4.7), (4.8) enables us to draw the following conclusions about the branching process $\xi(t)$, $t \geqslant 0$ itself.

In general we have a positive probability that no particle is left over after some time t. (Of course, this cannot happen if $\lambda_{10} = 0$, i.e., if the particles cannot vanish, but can only multiply.) If we have one particle at the initial moment $t = 0$, then this probability is $p_0(t) = F(t,0)$. If we are given k particles at the start, then this probability is $p_{k0}(t) = F(t,0)^k = p_0(t)^k$ [see (4.4)].

The function $p_0(t)$ is a solution of the differential equation (4.7) with the parameter $z = 0$;

$$p_0^!(t) = f(p_0(t)), \quad p_0(0) = 0.$$

We have seen that for $t \to \infty$, this solution tends asymptotically to the value α, which is the smallest root of the equation $f(x) = 0$ [see (4.9)], i.e.

$$\lim\limits_{t \to \infty} p_0(t) = \alpha .$$

We may say that α is the *probability of degeneration of the branching process* $\xi(t)$ -- the probability that after some period of time no particle is left over.

If we are given k particles at the initial time $t = 0$, then the probability of degeneration is equal to

(4.13) $\lim\limits_{t \to \infty} p_{k0}(t) = \alpha^k.$

Looking at the differential equations (4.7), (4.8), we see that with the condition (4.10) each particle generates with probability 1, a finite number of particles after a finite period of time, since,

(4.14)
$$\lim\limits_{z \to 1} F(t,z) = \sum\limits_{n} p_{1n}(t) = 1,$$
$$p_{1\infty}(t) = 1 - \sum\limits_{n} p_{1n}(t) \equiv 0, \quad t \geqslant 0.$$

When condition (4.11) holds, we have according to (4.12)

(4.15) $\lim\limits_{z \to 1} F(t,z) = x_0(t) = 1 - p_{1\infty}(t) < 1,$

where $x_0(t) < 1$ for $t > 0$, and, hence, one particle will, with positive probability $p_{1\infty}(t) > 0$, generate an infinite number of particles after a finite period of time t. If we look at equation (4.4), we can conclude that k particles existing at the initial time generate with probability

(4.16) $p_{k\infty}(t) = 1 - x_0(t)^k$

an infinite number of particles after a finite time t. (This phenomenon is called an *explosion*.)

We see that this "explosion" effect is an example of a condition, under which the differential equations (2.11) for the transition probabilities are violated. In fact, if $\lambda_{1\infty} = 0$, then according to (4.16) $\lambda_{k\infty} = p'_{k\infty}(0) = 0$, for all $k = 0,1, \ldots$; with density $\lambda_{\infty\infty} = p'_{\infty\infty}(0) = 0$ and $p_{\infty\infty}(t) \equiv 1$, equation (2.12) for $p_{1\infty}(t)$ will be

$$p'_{1\infty}(t) = \sum_k p_{1k}(t)\lambda_{k\infty} = 0, \quad p_{1\infty}(0) = 0.$$

Its unique solution is $p_{1\infty}(t) = 0$, which is a contradiction to (4.15).

Problem: Is an "explosion" possible for division into halves (cf. the problem on page 28)?

Section 5
Brownian Motion
The Diffusion Equation and Some Properties of the Trajectories

We consider a particle moving in a homogeneous fluid. It undergoes chaotic collisions with the molecules of the fluid, and as a result of this, it obeys a continuous disorder motion which is called Brownian motion.

As a discrete analogue of this process we can consider the following model of a *"random walk"*. Suppose we have a particle on the real line which is subject to the influence of external impulses from time to time, and which, after each impulse, moves to the value $\pm\Delta x$ (the direction depending on the direction of the impulse). Let the time interval between the different impulses be independent random variables which have identical exponential probability distributions with parameter λ. Let the displacements $\pm\Delta x$ be equally probable.

We suppose that the possible positions of the particles on the real line are the points $x = k\Delta x$, $k = 0,\pm 1, ...$, and we consider the position $\xi(t)$ of the particle at time t. According to what we have already said, the variables $\xi(t)$ will, in the course of time $t \geqslant 0$, form a homogeneous Markov process of the type that we considered in Section 1 (the states $x = k\Delta x$ can be enumerated in a natural way by the numbers $k = 0,\pm 1, ...$). Initially, the particle is at the point $\xi(0) = 0$; after a random time which has an exponential distribution with parameter λ, it moves with equal probability to one of the states $\pm\Delta x$; if the particle has reached the point x, it stays there independently of its past behaviour during a random time τ, which is exponentially distributed with parameter λ, and then moves with the same probability to one of the states $x\pm\Delta x$. More precisely, we can say that the time τ considered is the distance between subsequent impulses which produce the shift of the particle; the mean value $M\tau$ of this time interval is equal to $1/\lambda$.

We look at the result of this model of a random walk with step width Δx in the limit $\Delta x \to 0$ and $\lambda \to \infty$. We consider the transition probability of our process, denoting by $p(x,t,y)$ the probability of transition from the point x to the point y during the period of time t:

$$p(x,t,y) = \mathbf{P}\{\xi(s + t) = y \mid \xi(s) = x\}$$

(for $x = i\Delta x$ and $y = j\Delta x$ this is nothing other than $p_{ij}(t)$ in our previous notation).

According to the backward system of differential equations (2.10) with parameters

$$\lambda_{i,i-1} = \lambda_{i,i+1} = \frac{1}{2}, \quad \lambda_{ii} = -\lambda ,$$

$$\lambda_{ij} = 0 \quad \text{for} \quad j \neq i - 1, i, i + 1$$

the transition probability $p(x,t,y)$ satisfies, as a function of t and x, the differential equation

$$\frac{\partial}{\partial t} p(x,t,y) = \frac{1}{2} \lambda[p(x + \Delta x, t, y) - 2p(x,t,y) + p(x - \Delta x, t,y)].$$

We introduce the function

$$f(x,t) = \sum_{y} \varphi(y)p(x,t,y), \quad t \geqslant 0,$$

this is the mean value of the random variable $\varphi(\xi(s + t))$ for a given function $\varphi(y)$, $-\infty < y < \infty$, and $\xi(s) = x$:

(5.1) $$f(x,t) = \mathbf{M}_{s,x}\varphi(\xi(s + t)),$$

where $\mathbf{M}_{s,x}$ denotes the conditional mathematical expectation under the condition $\xi(s) = x$; it is obvious that

(5.2) $$f(x,0) = \varphi(x), \quad -\infty < x < \infty .$$

For a function $\varphi(y)$, which is zero outside some interval $|y| \leqslant C$, the function $f(x,t)$ satisfies the differential difference equation

$$\frac{\partial}{\partial t} f(x,t) = \frac{1}{2}\lambda[f(x + \Delta x, t) - 2f(x,t) + f(x - \Delta x, t)].$$

We obtain this equation from the differential difference equation above if we multiply $p(t,x,y)$ by $\varphi(y)$ and take the sum over the finitely many $y = k\Delta x$, $|y| \leqslant C$.

We set $\lambda(\Delta x)^2 = \sigma^2$, where $\sigma^2 > 0$ is some constant, and we write the equation in the form

$$\frac{\partial}{\partial t} f(x,t) = \frac{1}{2} \sigma^2 \frac{f(x + \Delta x,t) - 2f(x,t) + f(x - \Delta x,t)}{(\Delta x)^2} .$$

To what does this equation tend for $\Delta x \to 0$?

If we have a twice continuous differentiable function $f(x)$, then we know that

$$\lim_{\Delta x \to 0} \frac{f(x + \Delta x) - 2f(x) + f(x - \Delta x)}{(\Delta x)^2} = \frac{d^2}{dx^2} f(x),$$

and, hence, we obtain for $\Delta x \to 0$ the limit equation

(5.3) $$\frac{\partial}{\partial t} f(x,t) = \frac{1}{2} \sigma^2 \frac{\partial^2}{\partial x^2} f(x,t).$$

We use this result to construct a model of Brownian motion as a continuous random walk $\xi(t)$, $t \geqslant 0$, for which the solution $f(x,t)$ with the initial condition (5.2) can be interpreted in the same way as (5.1).

The equation (5.3) is the well-known *diffusion equation* (the constant σ^2 denotes the *diffusion coefficient*), which has the solution

(5.4) $$f(x,t) = \int_{-\infty}^{\infty} \varphi(y) p(x,t,y) dy,$$

where

(5.5) $$p(x,t,y) = \frac{1}{(2\pi\sigma^2 t)^{1/2}} e^{-(y-x)^2/2\sigma^2 t}, \quad -\infty < y < \infty.$$

We see immediately that the function $p(x,t,y)$, which we call the *fundamental solution* of equation (5.3), is the density of the *normal distribution* with mean value x and variance $\sigma^2 t$.

We get a continuous model of Brownian motion, where the position of the particle at time t, described by the random variable $\xi(t)$, changes during the time t, from the following prescription: For any times $s_1, \ldots < s_m < s < t$, the conditional density of distribution of $\xi(t)$, given that

$$\xi(s_1) = x_1, \ldots, \xi(s_m) = x_m, \quad \xi(s) = x,$$

exists and is defined by

(5.6) $$p(x, t - s, y) = \frac{1}{(2\pi\sigma^2(t-s))^{1/2}} e^{-(y-x)^2/2\sigma^2(t-s)}, \quad -\infty < y < \infty.$$

Given $\xi(s) = x$, the density (5.6) does not depend on the conditions

$$\xi(s_1) = x_1, \ldots, \xi(s_m) = x_m$$

in the "past". (This is the so-called *Markov property*), and moreover, they do not depend on the location of the interval (s,t) on the time axis. Recall that both the Markov property and the *homogeneity in time* observed above, were characteristic properties for the initial random walk with step width $\Delta x \to 0$.

Problem: Let $\xi(t)$, $t \geqslant 0$ be a family of random variables having the properties described above [see (5.6)], where

(5.7) $$\xi(0) = 0.$$

Show that for $s_1 < \ldots < s_m < s < t$ the variables $\xi(t) - \xi(s)$ do not depend on the entire past,

$$\xi(s_1), ..., \xi(s_m), \xi(s),$$

and that they have normal probability distribution with mean value zero and variance

(5.8) $\mathbf{M}[\xi(t) - \xi(s)]^2 = \sigma^2(t - s),$

and that in fact the density of the probability distribution of the variable $\xi(t) - \xi(s)$ is

(5.9) $p(0, t - s, y) = \dfrac{1}{(2\pi\sigma^2 (t-s))^{1/2}} e^{-y^2/2\sigma^2(t-s)}, \quad -\infty < y < \infty .$

Hint: Apply the fact that the density of the conditional probability distribution of the variable $\xi(t) - \xi(s)$, under the conditions $\xi(s_1) = x_1, ..., \xi(s_m) = x_m, \xi(s) = x$, is the same as the density of the conditional probability distribution of the variable $\xi(t) - x$, which is equal to $p(0, t - s, y)$, and does not depend on $x_1, ..., x_m, x$.

Problem: Show that for any $0 < t_1 < ... < t_n$, the variables $\xi(t_1), ..., \xi(t_n)$ have the joint distribution density

(5.10)
$$p_{t_1,...,t_n}(x_1, ..., x_n)$$
$$= p(0,t_1,x_1) \cdot p(x_1, t_2 - t_1, x_2)...p(x_{n-1}, t_n - t_{n-1}, x_n).$$

Show that the increments

(5.11) $\xi(t_1) - \xi(0), ..., \xi(t_n) - \xi(t_{n-1})$

are a family of independent random variables.

Let $\xi(t)$, $t \geqslant 0$ be random variables depending on the parameter t (time) which have the properties described in (5.6) - (5.11). It will be convenient to characterize the variables $\xi(t)$, $t \geqslant 0$, in the following way: $\xi(0) = 0$, and the increments

$$\xi(t_1) - \xi(0) = \xi(t_1), ..., \xi(t_n) - \xi(t_{n-1})$$

are independent for arbitrary $0 < t_1 < ... < t_n$, where for each interval (s,t) the increment $\xi(t) - \xi(s)$ is a random variable, having normal probability distribution with mean value 0 and variance $\sigma^2(t - s)$.

Problem: Show that the relations described in (5.6) hold if the property (5.11) is proved for $\xi(t)$, $t \geqslant 0$.

Let us look at the space of elementary events Ω, for which we define formally the random variables

$$\xi(t) = \xi(\omega, t), \qquad \omega \in \Omega$$

If we interpret $\xi(t)$ as the position of the Brownian particle at time t, then we can say that for the elementary outcome $\omega \in \Omega$ we have the corresponding trajectory

$$x(t) = \xi(\omega, t), \qquad t \geqslant 0.$$

We characterized the random variables $\xi(t)$, $t \geqslant 0$ from the point of view of their probability distribution [see (5.6) - (5.11)], and this characterization still remains somewhat arbitrary regarding the dependence of $\xi(t) = \xi(\omega, t)$ on $\omega \in \Omega$. For example, we may, without violating the properties expressed in (5.6) - (5.11), arbitrarily change the value $\xi(\omega, t)$ of all variables $\xi(t)$ for some event ω (of probability 0), and therefore the trajectory $x(t) = \xi(\omega, t)$, $t \geqslant 0$, is not uniquely defined. To circumvent this difficulty, we define a (random) trajectory with probability 1 of the motion of the Brownian particle by a sequence of approximations with respect to the position $\xi(t_{kn})$ at discrete times t_{kn}. As approximations we take *random variables which are piecewise linear*:

$$\xi_n(t) = \frac{t - t_{k+1,n}}{t_{kn} - t_{k+1,n}} \, \xi(t_{kn}) + \frac{t - t_{kn}}{t_{k+1,n} - t_{kn}} \, \xi(t_{k+1,n}),$$

(5.12)

$$t_{kn} \leqslant t \leqslant t_{k+1,n} \, .$$

Each nth trajectory subsequently joins by its linear segments the points $\xi(t_{kn})$, where

$$t_{kn} = \frac{k}{2^n} \, , \qquad k = 0, 1, \ldots \, .$$

Theorem. *The random function (5.12) converges uniformly on each finite time interval with probability 1.*

Proof: Let us consider the events

$$A_T^{m,n} = \left\{ \max_{0 \leqslant t \leqslant T} |\xi_n(t) - \xi_m(t)| > \epsilon_m \right\},$$

where $n > m$ and T is a positive integer, $T > 0$. It is obvious that for the function (5.12) the maximum that we consider here is attained at one of the vertex points $t_{kn} = k/2^n$; as n becomes larger, the maximum can only increase, since the vertex points t_{kn} that we have for smaller n and their values $\xi(t_{kn})$ there do not change if n becomes larger. For the union $A_T^m = \cup_{n>m} A_T^{m,n}$ of the monotone non-decreasing events $A_T^{m,n}$, $n = m + 1, m + 2, \ldots$, we have that

$$P(A_T^m) = \lim_{n \to \infty} P(A_T^{m,n}).$$

In the following, we shall obtain an estimation of the probability $P(A_T^{m,n})$, uniformly for all n, which will also be an estimation of the probability $P(A_T^m)$ of the event A_T^m, which says that

$$\max_{0 \leqslant t \leqslant T} |\xi_n(t) - \xi_m(t)| > \epsilon_m$$

for at least one $n > m$.

It is obvious that

$$P(A_T^{m,n}) \leqslant 2^m T \cdot P \left\{ \max_{0 \leqslant t \leqslant 2^{-m}} |\xi_n(t) - \xi_m(t)| > \epsilon_m \right\}$$

$$\leqslant 2^m T \cdot P \left\{ \max_{0 \leqslant k \leqslant 2^{n-m}} (|\xi(t_{kn})|, |\xi(t_{kn}) - \xi(2^{-m})|) > \epsilon_m \right\}$$

$$\leqslant 2^m T \cdot 4 P \left\{ \max_{0 \leqslant k \leqslant 2^{n-m}} \xi(t_{kn}) > \epsilon_m \right\}$$

(here we apply the fact that the family of variables $-\xi(t)$, $0 \leqslant t \leqslant h$, as well as the family of variables $\xi(h - t) - \xi(h)$, $0 \leqslant t \leqslant h$, have the same probability distribution for corresponding t as the variables $\xi(t)$, $0 \leqslant t \leqslant h$). It is possible to apply the following general theorem to our variables $\xi(t_{kn})$, $k = 1, ..., 2^{n-m}$.

Lemma. *Let the random variables $\xi_1, ..., \xi_n$ be such that for all $k = 1, ..., n-1$ the differences $\xi_n - \xi_k$ do not depend on $\xi_1, ..., \xi_k$, where their probability distribution on the real line is symmetric to the point 0. Then*

$$(5.13) \qquad P \left\{ \max_{1 \leqslant k \leqslant n} \xi_k > x \right\} \leqslant 2P\{\xi_n > x\}, \quad x > 0.$$

Proof: Under the condition that $\max_{1 \leqslant k \leqslant n} \xi_k > x$, and denoting by ξ_ν the first of the variables $\xi_1, ..., \xi_n$ that exceeds the level of x, and if we take into consideration that the event $\{\nu = k\}$ is determined by the first k variables $\xi_1, ..., \xi_k$, but that the difference $\xi_n - \xi_k$ does not depend on them, then we have

$$P \left\{ \max_{0 \leqslant k \leqslant n} \xi_k > x, \; \xi_n \leqslant x \right\}$$

$$= \sum_{k=0}^{n-1} P\{\nu = k, \; \xi_n \leqslant x\} \leqslant \sum_{k=0}^{n-1} P\{\nu = k, \; \xi_n - \xi_k < 0\}$$

$$= \sum_{k=0}^{n-1} P\{\nu = k\} \cdot P\{\xi_n - \xi_k < 0\}$$

$$\leqslant \sum_{k=0}^{n-1} P\{\nu = k\} \cdot P\{\xi_n - \xi_k \geqslant 0\}$$

$$= \sum_{k=0}^{n-1} P\{\nu = k, \; \xi_n - \xi_k \geqslant 0\} \leqslant P\{\xi_n > x\}.$$

If we add the inequality

$$P\left\{ \max_{0 \leqslant k \leqslant n} \xi_k > x, \ \xi_n > x \right\} \leqslant P\{\xi_n > x\}$$

we obtain our estimation (5.13). $\square$

Applying the general estimation (5.13) to the variables $\xi(t_{kn})$, we get

$$P\left\{ \max_{0 \leqslant k \leqslant 2^{n-m}} \xi(t_{kn}) > \epsilon_m \right\} \geqslant 2P\{\xi(2^{-m}) > \epsilon_m\},$$

where, for the variable $\xi(2^{-m})$, which is the last of the subsequent variables $\xi(t_{kn})$, $k = 1, ..., 2^{n-m}$, that we consider here, the estimation

$$P\{\xi(2^{-m}) > \epsilon_m\} = \frac{1}{\sqrt{2\pi}} \int_{\epsilon_m \sqrt{2^m}/\sigma} e^{-x^2/2} \, dx$$

$$\leqslant \frac{1}{\sqrt{2\pi}} \frac{\sigma}{\epsilon_m \sqrt{2^m}} \int_{\epsilon_m \sqrt{2^m}/\sigma} x e^{-x^2/2} dx$$

$$= \frac{1}{\sqrt{2\pi}} \frac{\sigma}{\epsilon_m \sqrt{2^m}} e^{-\epsilon_m^2 2^m/2\sigma^2}$$

holds. The final result is the following estimation:

$$P(A_T^m) \leqslant 4T\sigma \sqrt{2/\pi} \frac{\sqrt{2^m}}{\epsilon_m} e^{-\epsilon_m^2 2^m/2\sigma^2}.$$

We choose $\epsilon_m \to 0$ such that the series

$$\sum_{m=1}^{\infty} \frac{\sqrt{2^m}}{\epsilon_m} e^{-\epsilon_m^2 2^m/2\sigma^2} < \infty$$

converges (for example, we may take $\epsilon_m = 2^{-m/4}$). Then, from the convergence of the series

$$\sum_{m=1}^{\infty} P(A_T^m) < \infty,$$

using the well-known Borel-Cantelli lemma, we have the result that with probabability 1 among A_T^m, $m = 1,2, ...$, only a finite number of events occurs; this signifies that with probability 1, for sufficiently large m and for all $n > m$,

$$\max_{0 \leqslant t \leqslant T} |\xi_n(t) - \xi_m(t)| \leqslant \epsilon_m,$$

where $\epsilon_m \to 0$ for $m \to \infty$. We see that with probability 1 the sequence of random variables (5.12) converges uniformly on every finite interval $0 \leqslant t \leqslant T$, which we wanted to demonstrate. $\square$

We now come to the final definition of the initial variables $\xi(t)$, $t \geqslant 0$, as follows: for each elementary event $\omega \in \Omega$ for which the sequence of functions (5.12) converges uniformly on each finite interval $0 \leqslant t \leqslant T$ we set

(5.14) $\xi(\omega,t) = \lim_{n \to \infty} \xi_n(\omega,t), \quad t \geqslant 0.$

Formula (5.14) determines the random variables $\xi(t)$, $t \geqslant 0$, with probability 1 (for almost all elementary events $\omega \in \Omega$). The corresponding trajectory $x(t) = \xi(\omega,t)$, $t \geqslant 0$, is continuous, because it is the uniform limit of the continuous functions $x_n(t) = \xi_n(\omega,t)$, $t \geqslant 0$, on each finite interval.

Taking all the above into account, we give the final definitions. A family of random variables $\xi(t)$, $t \geqslant 0$, depending on the parameter t (time) is called a *random process*; the corresponding function

$$x(t) = \xi(\omega, t), \quad t \geqslant 0;$$

is called the *trajectory* for the elementary outcome ω; it depends on $\omega \in \Omega$ and is random in this sense. *Brownian motion* is the random process $\xi(t)$, $t \geqslant 0$, which has the properties (5.6) - (5.11); it is usually supposed that the trajectories are continuous with probability 1. Brownian motion is sometimes called the *Wiener process*. As one of its definitions from the point of view of the conditions described in (5.6) - (5.11) we could also take the following: $\xi(0) = 0$, the increment $\xi(t) - \xi(s)$ has normal distribution with expectation 0 and variance $\sigma^2(t - s)$ on each finite interval (s,t), where for any $0 < t_1 < ... < t_n$ the increments $\xi(t_1) - \xi(0)$, ..., $\xi(t_n) - \xi(t_{n-1})$ are independent.

Problem: Show that for the Brownian motion $\xi(t)$, $t \geqslant 0$, the variable

$$\tau_a = \min\{t: \xi(t) \geqslant a\}, \quad a > 0,$$

-- the time at which the point $x = a$ is attained -- has the probability distribution

(5.15) $P\{\tau_a \leqslant t\} = 2P\{\xi(t) \geqslant a\}, \quad t \geqslant 0,$

with the density (for $\sigma^2 = 1$)

$$p(t) = \frac{a}{\sqrt{2\pi}} t^{-3/2} e^{-a^2/2t}, \quad t > 0.$$

Hint: Apply the fact that the trajectories are continuous and that under the condition $\tau_a \leqslant t$, the Brownian particle will, in its motion with initial state $\xi(\tau_a) = a$, appear with equal probability on the right or the left of the point $x = a$ at time $t \geqslant \tau_a$.

Problem: Show that the Brownian particle hits any arbitrary point x, $-\infty < x < \infty$, sooner or later with probability 1.

(5.16) $P\left\{\max_{0\leqslant s\leqslant t}\xi(s)\geqslant x\right\} = 2P\{\xi(t)\geqslant x\},\quad x\geqslant 0,$

holds for Brownian motion and that the density of distribution of the maximum described here (with $\sigma^2 = 1$) is given by

$$p(x) = \sqrt{2/\pi t}\,e^{-x^2/2t},\quad x\geqslant 0.$$

Hint: Apply formula (5.15).

Problem: Suppose that the Brownian particle is at the point a at time t: $\xi(t) = a$. Show that after any period of time, which may be as small as

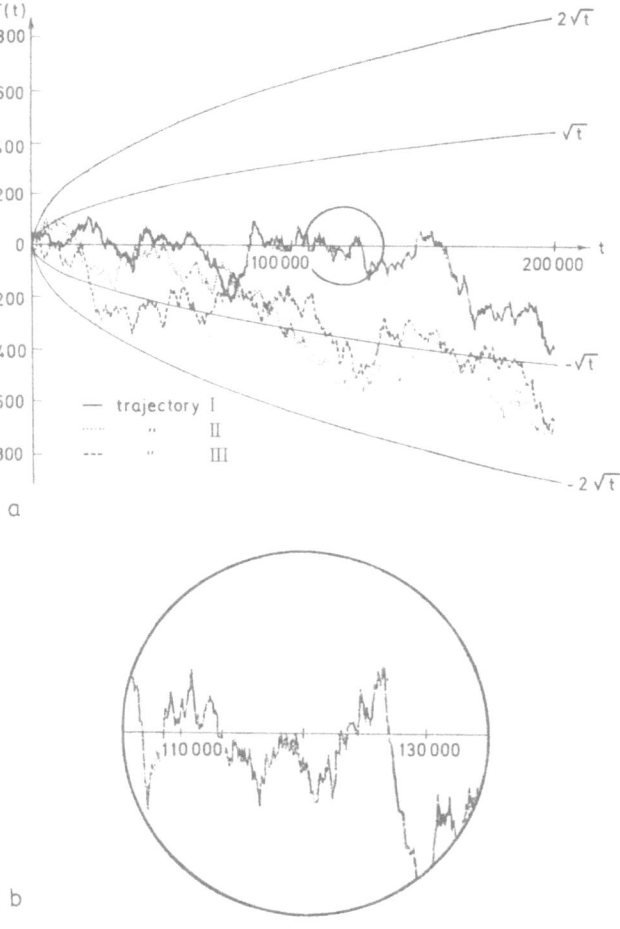

Fig. 6 a) Experimental trajectories of the Brownian motion with diffusion coefficient $\sigma^2 = 1$.
         b) part of the Figure 12 times enlarged.

one wants, the particle appears with probability 1 in the region $x < a$ as well as in the region $x > a$; more precisely, for any $h > 0$,

$$(5.17) \qquad P\left\{ \max_{t \leqslant s \leqslant t+h} \xi(s) > a, \quad \max_{t \leqslant s \leqslant t+h} \xi(s) < a | \xi(t) = a \right\} = 1$$

holds.

Some experimental trajectories of Brownian motion are drawn in Fig. 6[3].

The trajectories look as if they were chaotically drawn by a jittering pen (which reflects the character of the physical process of Brownian motion, where the particle is subject to infinitely frequent impulses from the molecules, and every impulse produces an infinitely small displacement). As we shall see below, the trajectory of the Brownian particle has unbounded variation on any interval with probability 1:

$$(5.18) \qquad \sup_{s=t_0<t_1<...<t_n=t} \sum_{k=1}^{n} |\xi(t_k) - \xi(t_{k-1})| = \infty .$$

Theorem. *For the process of Brownian motion, the following holds on any interval (s,t) with probability 1:*

$$(5.19) \qquad \lim_{n \to \infty} \sum_{k=1}^{n} [\xi(t_k) - \xi(t_{k-1})]^2 = \sigma^2(t - s),$$

where the limit is taken over a sequence of partitions $s = t_0 < t_1 ... < t_n = t$, $n = n_m$, with step width

$$h_m = \max_{1 \leqslant k \leqslant n} |t_{k,n} - t_{k-1,n}| \leqslant 2^{-m}.$$

Proof: Let us start by showing that the limit relation (5.19) holds for partitions with any wanted step width $h_m \to 0$, if we consider convergence in quadratic mean. In fact, we have

$$M[\xi(t_k) - \xi(t_{k-1})]^2 = \sigma^2(t_k - t_{k-1}).$$

If we set

$$\Delta_k = [\xi(t_k) - \xi(t_{k-1})]^2 - \sigma^2(t_k - t_{k-1})$$

---

[3]Cf. Bibliography of Time Series and Stochastic Processes (H. O. Wold, ed.), Edinburgh, London, 1965, pp. 10-11.

[4]For variables ξ which have normal distribution with expectation 0 and variance σ^2, the moment $M\xi^k$ can be easily found from the general formula

$$M(i\xi)^k = \frac{d^k}{du^k} Me^{iu\xi} \quad k = 0$$

using the characteristic function $Me^{iu\xi} = e^{-\sigma^2 u^2/2}$, $-\infty < u < \infty$.

and consider the sum

$$\sum_{k=1}^{n} [\xi(t_k) - \xi(t_{k-1})]^2 - \sigma^2(t - s) = \sum_{k=1}^{n} \Delta_k$$

of the independent variables Δ_k with expectation 0 and variance[4]

$$M\Delta_k^2 = M[\xi(t_k) - \xi(t_{k-1})]^4 - \sigma^4(t_k - t_{k-1})^2 = 2\sigma^4(t_k - t_{k-1})^2,$$

we obtain

$$M\left(\sum_{k=1}^{n} \Delta_k\right)^2 = \sum_{k=1}^{n} M\Delta_k^2 = 2\sigma^4 \cdot \sum_{k=1}^{n} (t_k - t_{k-1})^2$$

$$\leqslant 2\sigma^4 \cdot \max_{1 \leqslant k \leqslant n} (t_k - t_{k-1}) \cdot \sum_{k=1}^{n} (t_k - t_{k-1})$$

$$= 2\sigma^4 \cdot h_m(t - s) \to 0 \quad \text{with} \quad m \to \infty.$$

From the Cebyshev inequality,

$$P\left\{\left|\sum_{k=1}^{n} \Delta_k\right| > \epsilon_m\right\} \leqslant 2\sigma^4(t - s)\frac{h_m}{\epsilon_m^2}$$

holds and for $h_m \leqslant 2^{-m}$ we can choose $\epsilon_m \to 0$ such that

$$\sum_{m=1}^{\infty} P\left\{\left|\sum_{k=1}^{n} \Delta_k\right| > \epsilon_m\right\} < \infty.$$

Using the Borel-Cantelli lemma, only a finite number of events $\{\sum_{k=1}^{n}\Delta_k > \epsilon_m\}$ occurs with probability 1, i.e. for sufficiently large n and with probability 1 we have

$$\left|\sum_{k=1}^{n} \Delta_k\right| \leqslant \epsilon_m, \quad \text{where} \quad \epsilon_m \to 0.$$

The theorem is proved.   □

Problem: Show that the Brownian trajectory has unbounded variation on any interval with probability 1.

Hint: Apply the limit relation (5.19).

From now on, we shall call the process of Brownian motion $\xi(t), t \geqslant 0$, with diffusion coefficient $\sigma^2 = 1$, *standard* Brownian motion. This process plays a very important role in the whole theory of random processes and it is the basis of many probabilistic theoretical models. We shall often be dealing with the *standard process of Brownian motion* (sometimes called *standard Wiener process*)

$$\eta(t) = \xi(t - t_0), \quad t \geqslant t_0,$$

on the half-axis $t \geqslant t_0$.

Section 6
Random Processes in Multi-Server Systems

We shall consider two examples of random processes arising in different multi-server systems.
Let

(6.1) $\quad\quad \xi_0, \xi_1, ...$

be a sequence of positive independent, identically distributed random variables and

$$\tau_n = \xi_0 + \cdots + \xi_{n-1}, \quad n = 1, 2, ... \;.$$

Consider the case that we have an instrument which works without failure for a random period of time ξ_0. On failure after the random time ξ_0, it is replaced by a new instrument, which in turn fails after the random time ξ_1, being replaced by another new instrument and so on. Apart from the variable τ_n, it could be interesting to consider for *this process of replacements* for example the variables $\nu(t)$, i.e. the *number of replacements* during the time interval $[0,t]$ and $\Delta(t) = \tau_{\nu(t)+1} - t$, i.e. the remaining service time of the instrument that works at time t.

We suppose that the variables (6.1) are exponentially distributed with parameter λ; recall that this is the distribution with density

$$p(t) = \begin{cases} \lambda e^{-\lambda t}, & t \geqslant 0, \\[2mm] 0, & t < 0. \end{cases}$$

We show that the corresponding variable τ_n has the density

$$(6.2) \qquad p_n(t) = \begin{cases} \lambda \dfrac{(\lambda t)^{n-1}}{(n-1)!} e^{-\lambda t}, & t \geqslant 0, \\ 0 & , \quad t < 0. \end{cases}$$

In fact this holds for $n = 1$. If it is true for $n = k - 1 \geqslant 1$, then by applying the conditional density $p_n(t|\tau_{n-1} = s) = p(t - s)$, $t \geqslant s$ of the variable $\tau_n = \tau_{n-1} + \xi_n$ under the condition $\tau_{n-1} = s$, we get for $n = k$ that

$$p_n(t) = \int_0^t p(t - s) p_{n-1}(s) ds = \int_0^t \lambda e^{-\lambda(t-s)} \lambda \frac{(\lambda s)^{n-2}}{(n-2)!} e^{-\lambda s} ds$$

$$= \lambda e^{-\lambda t} \frac{\lambda^{n-1}}{(n-2)!} \int_0^t s^{n-2} ds = \lambda e^{-\lambda t} \frac{(\lambda t)^{n-1}}{(n-1)!}.$$

Problem: If the working time of one single instrument is exponentially distributed with parameter λ, then show that $\nu(t)$, $t \geqslant 0$, is a Poisson process with the same parameter

$$P\{\nu(t) = k\} = \frac{(\lambda t)^k}{k!} e^{-\lambda t}, \quad k = 0, 1, \ldots,$$

and the variable $\Delta(t)$ has the same exponential distribution:

$$P\{\Delta(t) > h\} = e^{-\lambda h}, \quad h \geqslant 0.$$

Now we interpret

$$\tau_1 = \xi_0, \quad \tau_2 = \xi_0 + \xi_1, \ldots$$

as a sequence of time intervals that corresponds to a sequence of "demands" made on our "service system". Let the immediate "service" of the nth demand take some random time η_n, independently of τ_1, τ_2, ..., and let

$$(6.3) \qquad \eta_1, \eta_2, \ldots$$

be a sequence of independent random variables with the same probability distribution.

In the *multi-server system* described, we shall be interested in the variable $\mathcal{H}_n$, i.e., the waiting time until the beginning of the service of the nth demand. In our notation, the time up to the end of the service of the nth demand is then $\mathcal{H}_n + \eta_n$. We suppose that the service system is vacant if the following $(n+1)$th demand comes in after the time $\xi_n \geqslant \mathcal{H}_n + \eta_n$, and the service starts immediately, that is, $\mathcal{H}_{n+1} = 0$. If $\xi_n < \mathcal{H}_n + \eta_n$, the system is still occupied by the previous demand at time $\tau_{n+1} = \tau_n + \xi_n$, the $(n+1)$th demand being forced to wait for the time period $\mathcal{H}_{n+1} = \mathcal{H}_n + \eta_n - \xi_n$, before it is served.

We set

(6.4) $\Delta_n = \eta_n - \xi_n, \quad S_n = \sum\limits_{k=1}^{n} \Delta_k, \quad n = 1,2, \dots$.

The relationship between the variables $\aleph_n$ and the independent random variables Δ_n, $n = 1,2, \dots$, is given by

(6.5) $\aleph_1 = 0, \quad \aleph_{n+1} = \begin{cases} 0 & , \quad \aleph_n + \Delta_n \leqslant 0, \\ \aleph_n + \Delta_n, & \aleph_n + \Delta_n > 0. \end{cases}$

Let us compare the sequence $\aleph_1$, $\aleph_2$ ... with the sequence $S_1, S_2, \dots$.
The first incoming demand to our system is immediately satisfied, thus $\aleph_1 = 0$. Obviously, $\aleph_2 = 0$, if $S_1 \leqslant 0$ and $\aleph_2 = S_1$, if $S_1 > 0$; we can express the relationship between $\aleph_2$ and S_1 as

$\aleph_2 = S_1 - \min(0, S_1).$

We suppose that

(6.6) $\aleph_{n+1} = S_n - \min(0, S_1, \dots, S_n).$

As we already said above, the $(n+1)$th demand occupies the service system during the time $\aleph_{n+1} + \eta_{n+1}$, and the following demand is required to wait for the time $\aleph_{n+2} = 0$, if $\aleph_{n+1} + \Delta_{n+1} \leqslant 0$, or $\aleph_{n+2} = \aleph_{n+1} + \Delta_{n+1}$, if $\aleph_{n+1} + \Delta_{n+1} > 0$. If we add the variable Δ_{n+1} on both sides of the equation (6.6), in the first case we get

$\aleph_{n+1} + \Delta_{n+1} = S_{n+1} - \min(0, S_1, \dots, S_n) \leqslant 0.$

Obviously,

$S_{n+1} = \min(0, S_1, \dots, S_{n+1}),$

and we find that

$0 = \aleph_{n+2} = S_{n+1} - \min(0, S_1, \dots, S_{n+1}).$

If $\aleph_{n+1} + \Delta_{n+1} > 0$, we get

$\aleph_{n+2} = \aleph_{n+1} + \Delta_{n+1} = S_{n+1} - \min(0, S_1, \dots, S_n) > 0,$

where obviously for $S_{n+1} > 0$

$\min(0, S_1, \dots, S_n) = \min(0, S_1, \dots, S_{n+1}).$

We can observe that formula (6.6) still holds, if we replace n by $n + 1$, and, hence, this formula is valid for all $n = 1,2, \dots$.

Let us now consider the following sums of these independent, identically distributed variables Δ_1, ..., Δ_n, reversing the summation order

$$S_1' = \Delta_n, \; S_2' = \Delta_n + \Delta_{n-1}, \; ..., \; S_n' = \Delta_n + \cdots + \Delta_1.$$

Obviously, the probability distribution of the variables $(S_1', ..., S_n')$ is the same as for the variables $(S_1, ..., S_n)$, that is

$$\max(0, S_1', ..., S_n') = \max(S_n - 0, S_n - S_1, ..., S_n - S_n)$$

$$= S_n - \min(0, S_1, ..., S_n).$$

This leads to the conclusion:

Theorem. *The probability distribution of the variable* $\aleph_{n+1}$ *is the same as the distribution of the variable*

$$\zeta_n = \max(0, S_1, ..., S_n);$$

in particular, for arbitrary $t \geqslant 0$,

(6.7) $P\{\aleph_{n+1} \leqslant t\} = P\{\zeta_n \leqslant t\}.$

Let us examine the behaviour of our multi-server system for large n, more exactly for $n \to \infty$.

Problem: Assume that the average service time exceeds the average time interval between successive demands, i.e.,

(6.8) $a = M\Delta_1 = M\eta_1 - M\xi_1 > 0.$

Show that $\aleph_n \to \infty$ for $n \to \infty$, where we consider convergence in probability. That is, for arbitrarily large t

(6.9) $P\{\aleph_n \geqslant t\} \to 1.$

Hint: Apply the law of large numbers for independent, identically distributed random variables $\Delta_1, \Delta_2, ...,$ according to which $S_n/n \to a > 0$.

Problem: Let

(6.10) $a = M\Delta_1 = M\eta_1 - M\xi_1 < 0.$

Given the sequence of sums (6.4), show that with probability 1 only a finite number of variables S_n is positive, and, hence, the variable

(6.11) $\zeta = \lim_{n \to \infty} \zeta_n = \max(0, S_1, ..., S_n, ...)$

is finite.

Hint: Apply the strong law of large numbers, according to which

$$\frac{S_n}{n} \to a < 0.$$

with probability 1.

Assume that the variables ζ_n in (6.11) increase monotonically, which implies that for arbitrary $t \geqslant 0$

$$P\{\zeta \leqslant t\} = \lim_{n \to \infty} P\{\zeta_n \leqslant t\}$$

is the limit of the monotonically decreasing probabilities $P\{\zeta_n \leqslant t\}$. According to the general formula (6.7), the limit distribution is

(6.12) $\lim_{n \to \infty} P\{\aleph_n \leqslant t\} = P\{\zeta \leqslant t\}, \quad t \geqslant 0.$

Let us find the limit distribution, if the variables (6.1) and (6.3) are exponentially distributed with respective parameters λ and μ. We can immediately say that the task isn't easy at all.

We shall need the distribution density of the variable $\Delta_1 = \eta_1 - \zeta_1$, given by

(6.13) $p(x) = \begin{cases} \dfrac{\mu}{\lambda + \mu} \lambda e^{\lambda x}, & x < 0, \\[3mm] \dfrac{\lambda}{\lambda + \mu} \mu e^{-\mu x}, & x > 0 . \end{cases}$

(Verify this! Recall that ζ_1 and η_1 are independent random variables, exponentially distributed with parameter λ and μ).

Assume we have found the distribution function $F_\zeta(x) = P\{\zeta \leqslant x\}$, described on the right hand side of (6.12); we shall see that it fulfills the integral equation

(6.14) $F_\zeta(x) = \int_{-\infty}^{x} F_\zeta(x - y)p(y)dy.$

In fact, the maximum $\widetilde{\zeta} = \max(0, \widetilde{S}_1, ..., \widetilde{S}_n, ...)$ of the variables $\widetilde{S}_n = \sum_{k=2}^{n+1}\Delta_k$ has the same probability distribution as the maximum ζ, where

$$P\{\zeta \leqslant x\} = P\{\Delta_1 \leqslant x, \Delta_1 + \widetilde{\zeta} \leqslant x\}.$$

Applying the formula of complete probability, where

$$F_\zeta(x - y) = P\{\Delta_1 + \widetilde{\zeta} \leqslant x | \Delta_1 = y\}$$

is the conditional probability under the condition $\Delta_1 = y$, we obtain (6.14).

Let S_1^+ be the first positive sum in the sequence S_n, $n = 1,2, ...,$ setting $S_1^+ = 0$, if $S_n \leqslant 0$, for all n. We can derive the probability distribution of the variable S_1^+ as follows: Obviously, for $x \geqslant 0$

$$P\{S_1^+ > x\} = \sum_{n=1}^{\infty} P\{S_1^+ > x, S_1^+ = S_n\}$$

$$= P\{S_1 > x\} + \sum_{n=1}^{\infty} P\{S_{n+1} > x, S_1 \leqslant 0, ..., S_n \leqslant 0\},$$

where $S_1 = \Delta_1$, $S_{n+1} = \Delta_{n+1} + S_n$ and

$$P\{\Delta_1 > x\} = \int_x^{\infty} \frac{\lambda}{\lambda + \mu} \mu e^{-\mu y} dy = \frac{\lambda}{\lambda + \mu} e^{-\mu x} = C_1 e^{-\mu x},$$

$$P\{\Delta_{n+1} > x - S_n, S_1 \leqslant 0, ..., S_n \leqslant 0\}$$

$$= \int_{-\infty}^0 \cdots \int_{-\infty}^0 \frac{\lambda}{\lambda + \mu} e^{-\mu(x - y_n)} P_{S_1,...,S_n}(dy_1 ... dy_n)$$

$$= C_{n+1} e^{-\mu x} ;$$

Here we make use of the fact that Δ_{n+1} is independent of $S_1, ..., S_n$. Hence,

(6.15) $P\{S_1^+ > x\} = p^+ e^{-\mu x}$,   for $x \geqslant 0$

where the constant $p^+ = \sum_{n=1}^{\infty} C_n$ has a simple probabilistic interpretation:

(6.16) $p^+ = P\{S_1^+ > 0\}$, $1 - p^+ = P\{S_1^+ = 0\}$.

We show that under condition (6.10)

(6.17) $p = 1 - p^+ > 0$.

According to the definition of the variable S_1^+, the maximum (6.11) can be found as

(6.18) $P\{\zeta = 0\} = p$.

In order to prove (6.17), we apply equation (6.14). For $p = 0$, we obtain the equation

$$0 = \int_{-\infty}^0 F_\zeta(-y) p(y) dy,$$

where

$$p(y) = \frac{\mu}{\lambda + \mu} \lambda e^{\lambda y} > 0.$$

This holds only if $F_\zeta(-y) = 0$, $y < 0$, but this condition contradicts condition (6.10), under which, as we know, the variable ζ is bounded and $F_\zeta(-y) \to 1$ for $y \to -\infty$.

Above, we introduced the variable $\Delta_1^+ = S_1^+$, which is equal to the first positive sum in the sequence

$$S_1 = \Delta_1, \quad S_2 = \Delta_1 + \Delta_2, \quad ...$$

and which is equal to zero, if such a sum does not appear. Suppose that such a sum exists and that $\Delta_1^+ = S_{n_1}$. Let us consider the sequence

$$S_1^{(1)} = \Delta_{n_1+1}, \quad S_2^{(1)} = \Delta_{n_1-1} + \Delta_{n_1+2}, \quad ...$$

of the sums of our variables Δ_k, beginning with the number $k = n_1 + 1$. These variables Δ_k are independent of $\Delta_1, ..., \Delta_{n_1}$ and Δ_1^+. We define Δ_2^+ as the first positive sum in our new sequence (say $\Delta_2^+ = S_{n_2}^{(1)}$), setting $\Delta_2^+ = 0$, if such a sum doesn't exist. It is clear that, if $\Delta_1^+ > 0$, the variable Δ_2^+ has the same probability distribution as the variable Δ_1^+. Under the conditions $\Delta_1^+ > 0$, $\Delta_2^+ > 0$ (if $\Delta_2^+ = S_{n_2}^{(1)}$), we define for the sequence

$$S_1^{(2)} = \Delta_{n_1+n_2+1}, \quad S_2^{(2)} = \Delta_{n_1+n_2+1} + \Delta_{n_1+n_2+2}, \quad ...$$

in a similar way the variable Δ_3^+ and so on. For each of the variables $\Delta_1^+, ..., \Delta_n^+$ defined above, we define, as above, for $\Delta_1^+ > 0$, ..., $\Delta_n^+ > 0$ the variable Δ_{n+1}^+, which is independent of $\Delta_1^+ > 0$, ..., $\Delta_n^+ > 0$ and has the same probability distribution as the variable Δ_1^+.

Under the condition $\Delta_1^+ > 0$, ..., $\Delta_n^+ > 0$, we set

$$S_n^+ = \sum_{k=1}^n \Delta_k^+, \quad n = 1,2, ... \ .$$

Obviously, for the maximum (6.11) with $x > 0$ we get

$$P\{0 < \zeta \leqslant x\} = P\{\Delta_1^+ > 0, \ S_1^+ \leqslant x, \ \Delta_2^+ = 0\}$$

$$+ P\{\Delta_1^+ > 0, \ \Delta_2^+ > 0, \ S_2^+ \leqslant x, \ \Delta_3^+ = 0\} + ...$$

$$+ P\{\Delta_1^+ > 0, \ ..., \ \Delta_n^+ > 0, \ S_n^+ \leqslant x, \ \Delta_{n+1}^+ = 0\} + ... \ .$$

According to formula (6.15), which we obtained above, and under the condition $\Delta_1^+ > 0$, ..., $\Delta_n^+ > 0$, the variable S_n^+ is the sum of the independent variables $\Delta_1^+, ..., \Delta_n^+$, having identically the same exponential probability distribution with parameter μ, and the variable Δ_{n+1}^+ does not depend on $\Delta_1^+, ..., \Delta_n^+$, so that

$$P\{S_n^+ \leqslant x, \ \Delta_{n+1}^+ = 0 \mid \Delta_1^+ > 0, \ ..., \ \Delta_n^+ > 0\} = p \cdot \int_0^x p_n(y)dy,$$

where

$$p = P\{\Delta_{n+1}^+ = 0 | \Delta_1^+ > 0, \ldots, \Delta_n^+ > 0\} = P\{\Delta_1^+ = 0\} = P\{\zeta = 0\},$$

$$p_n(y) = \mu \frac{(\mu y)^{n-1}}{(n-1)!} e^{-\mu y}, \quad y \geqslant 0$$

(cf. the general formula (6.2)). Using

$$P\{\Delta_1^+ > 0, \ldots, \Delta_n^+ > 0\} = (1 - p)^n,$$

we obtain the following result:

$$P\{0 < \zeta \leqslant x\} = \int_0^x \left[\sum_{n=1}^\infty p(1-p)^n \mu \frac{(\mu y)^{n-1}}{(n-1)!} \right] e^{-\mu y} dy$$

$$= p(1-p)\mu \int_0^x e^{-p\mu y} dy = (1-p)(1 - e^{-p\mu x}).$$

Obviously, for $\zeta > 0$, the variable ζ is exponentially distributed with a parameter equal to the product $p\mu$, where, as you will recall, $p = P\{\zeta = 0\} > 0$ (cf. (6.17), (6.18)). It is possible to determine the probability p, still unknown, from equation (6.14) applied to the distribution function

(6.19) $F_\zeta(x) = p + (1-p)(1 - e^{-p\mu x}), \quad x \geqslant 0,$

taking in this equation $x = 0$.

Let us formulate our result with regard to the distribution of the maximum (6.11) which, according to (6.12), is the limit distribution for $n \to \infty$ of the waiting time $\mathcal{K}_n$ in which we are interested.

Theorem. *The limit distribution (6.12) for the waiting time $\mathcal{K}_n$, $n \to \infty$ is given by formula (6.19).*

Accordingly, we can conclude that, if the system is working a long time, the incoming demand finds the system vacant with probability $p > 0$, and in the other case, the waiting time before the beginning of the service is exponentially distributed with parameter $p\mu$.

Problem: Show that this result holds for an arbitrary probability distribution of the variable ξ_1, which is the time interval between the incoming demands (of course, under condition (6.10)).

Problem: If the time interval between successive demands is exponentially distributed (with parameter λ), prove that condition (6.10) implies that $\lambda < \mu$ and that the limit probability $p = \lim_{n\to\infty} P\{\mathcal{K}_n = 0\} = P\{\zeta = 0\}$ of the event, that the following demand finds the system vacant, is

(6.20) $p = 1 - \dfrac{\lambda}{\mu}.$

Section 7
Random Processes as Functions in Hilbert Space

One of the most useful approaches to the study of random processes is, as we shall see in the following, the introduction of the Hilbert space H of random variables ξ, $M|\xi|^2 < \infty$, with *scalar product*

$$(7.1) \qquad (\xi_1,\xi_2) = M\xi_1\bar{\xi}_2$$

and the *norm of quadratic mean*

$$(7.2) \qquad \|\xi\| = (M|\xi|^2)^{1/2}.$$

This space is *complete* and, hence, *each fundamental sequence of variables* $\xi_n \in H$:

$$(7.3) \qquad \|\xi_n - \xi_m\| \to 0$$

has a limit in quadratic mean $\xi = \lim_{n\to\infty}\xi_n$ *in the space* H *for* $n,m \to \infty$; *i.e., there exists a variable* $\xi \in H$ *such that*

$$\|\xi_n - \xi\| \to 0 \quad \text{for} \quad n \to \infty.$$

(In this space H, we identify variables that are equal with probability 1).

Obviously, for (real) random variables ξ_1, ξ_2 with mean value 0 ($M\xi_1 = M\xi_2 = 0$), the scalar product (7.1) expresses their *correlation*. As we shall see, it is possible to describe very clearly by the help of the Hilbert space H the conditional mathematical expectations, the conditional probabilities and other important characteristics of random variables.

We recall general properties of the scalar product (7.1):

$$(\xi, \xi) = M|\xi|^2 \geq 0,$$

and we have equality, if and only if, $\xi = 0$ with probability 1;

$$\sum_{k,j=1}^{n} c_k \overline{c}_j (\xi_k, \xi_j) = M \left| \sum_{k=1}^{n} c_k \xi_k \right|^2 \geq 0$$

for arbitrary $\xi_1, ..., \xi_n \in H$ and constants $c_1, ..., c_n$, which implies the inequality

$$|(\xi_1, \xi_2)| \leq \|\xi_1\| \|\xi_2\|.$$

This is well known for the mathematical expectation, for which we have

(7.4) $M|\xi_1, \xi_2| \leq (M|\xi_1|^2)^{1/2} (M|\xi_2|^2)^{1/2}.$

If we talk about a *random process*, we think of a process that is described by a function $\xi(t)$, $t \in T$, of the real variable t (time), running through some set $T \subseteq R^1$ on the real line, the values of this function being the random variables $\xi(t)$, which describe the state of the process at time t.

If $M|\xi(t)|^2 < \infty$, it is possible to conceive of $\xi(t)$, $t \in T$, as a function in the Hilbert space H; more precisely, as a function with values $\xi(t) \in H$. This is what we shall do from now on.

We interpret the *continuity in quadratic mean* and the *differentiability* of a random process $\xi(t)$, $t \in T$, as the property of a function in the Hilbert space H equipped with the norm of quadratic mean (7.2). For example, for a function $\xi(t)$, which is defined in a neighbourhood of the point s, *continuity* for $t = s$ means that

(7.5) $\lim_{t \to s} \|\xi(t) - \xi(s)\| = 0,$

and *differentiability* in this point means the existence of a variable $\xi'(s) \in H$, such that

(7.6) $\lim_{t \to s} \left\| \frac{\xi(t) - \xi(s)}{t - s} - \xi'(s) \right\| = 0.$

Problem: Show that the Brownian motion process $\xi(t)$, $t \geq 0$, is continuous in quadratic mean, but is not differentiable.

Hint: Apply the equation

$$\|\xi(t) - \xi(s)\|^2 = \sigma^2 |t - s|$$

(cf. (5.9)).

Problem: Let $\xi(t)$, $t \geq 0$, be a Poisson process the trajectories of which are step functions with jumps at the random points $\tau = \tau_1, \tau_2, ...$ (cf. Fig. 1 on pg. 4). Show that this random process is continuous in quadratic mean.

Hint: Apply that for a Poisson process with parameter λ, the equation

$$\|\xi(t) - \xi(s) - \lambda(t - s)\|^2 = \lambda|t - s|$$

holds.

Problem: Let $\xi(t)$, $t \in T$, be a random process with mean value $M\xi(t) = 0$ and *correlation function*

$$(7.7) \qquad B(t_1, t_2) = M\xi(t_1)\overline{\xi(t_2)} = (\xi(t_1), \xi(t_2)).$$

This is the name of this scalar product considered as a function of the variables $t_1, t_2 \in T$.

If the function $B(t_1, t_2)$ has the continuous second derivatives $\partial^2/\partial t_1^2$, $\partial^2/\partial t_1 \partial t_2$, $\partial^2/\partial t_2^2$ in a neighbourhood of the point $t_1 = s$, $t_2 = s$, then show that the random process $\xi(t)$ is continuously differentiable (in quadratic mean) in the neighbourhood of the point $t = s$, where

$$M\xi'(t_1)\overline{\xi'(t_2)} = \frac{\partial^2}{\partial t_1 \partial t_2} B(t_1, t_2).$$

We define the *integral* $\int_T \xi(t)dt$ for the random process $\xi(t)$, $t \in T$, as a function in the Hilbert space H on a finite or infinite interval T, beginning with the piecewise constant functions $\xi(t)$, taking only a finite number of values $\xi_k \in H$ different from 0, on nonintersecting half-intervals of the form $\Delta_k = (s_k, t_k] \subseteq T$:

$$(7.8) \qquad \xi(t) = \xi_k, \quad t \in \Delta_k.$$

For such functions $\xi(t)$, $t \in T$, we define their integral as

$$(7.9) \qquad \int_T \xi(t)dt = \sum_k \xi_k |\Delta_k|,$$

where $|\Delta| = t - s$ for the half-interval $\Delta = (s, t]$. Obviously, for arbitrary functions $\xi_1(t)$, $\xi_2(t)$ of the type (7.8) and constants c_1, c_2, the linear combination $\xi(t) = c_1\xi_1(t) + c_2\xi_2(t)$ is a function of the same type and

$$(7.10) \qquad \int_T (c_1\xi_1(t) + c_2\xi_2(t))dt = c_1\int_T \xi_1(t)dt + c_2\int_T \xi_2(t)dt.$$

The following relations are also obvious:

$$(7.11) \qquad \left\|\int_T \xi(t)dt\right\| \leq \int_T \|\xi(t)\|dt;$$

For arbitrary variables $\eta \in H$

$$(7.12) \qquad \left[\eta, \int_T \xi(t)dt\right] = \int_T (\eta, \xi(t))dt$$

and

(7.13) $\left[\int_T \xi_1(s)ds, \int_T \xi_2(t)dt \right] = \int_T \int_T (\xi_1(s), \xi_2(t))ds\,dt.$

Here we deal with the norm and the scalar product in the Hilbert space H (cf. (7.1), (7.2)).

An arbitrary function $\xi(t)$ is called *integrable in quadratic mean*, if there exists a sequence of piecewise constant functions $\xi_n(t)$ that approximate $\xi(t)$ in the sense that

(7.14) $\lim_{n \to 0} \int_T \| \xi(t) - \xi_n(t) \| dt = 0.$

Given an integrable function $\xi(t)$, the *integral* is defined as

(7.15) $\int_T \xi(t)dt = \lim_{n \to \infty} \int_T \xi_n(t)dt.$

This limit (in quadratic mean) does exist, since the sequence of integrals $\int_T \xi_n(t)dt$ is fundamental in the Hilbert space H, and, hence,

$$\left\| \int_T \xi_n(t)dt - \int_T \xi_m(t)dt \right\| = \left\| \int_T (\xi_n(t) - \xi_m(t))dt \right\|$$

$$\leqslant \int_T \| \xi_n(t) - \xi_m(t) \| dt$$

$$\leqslant \int_T \| \xi_n(t) - \xi(t) \| dt + \int_T \| \xi(t) - \xi_m(t) \| dt \to 0$$

for $n,m \to \infty$. Obviously, the limit in (7.15) does not depend on the choice of the approximating sequence $\xi_n(t)$, (verify this!).

Problem: Show that the relations (7.10) - (7.13) can be extended to any functions that are integrable in quadratic mean.

Problem: Let $\xi(t)$, $t \geqslant t_0$, be a random function which is continuous in quadratic mean. Show that it is integrable on any finite interval $T = [t_0, t]$ and that

$$\int_T \xi(s)ds = \int_{t_0}^t \xi(s)ds = \lim_{n \to \infty} \sum_{k=1}^{n} \xi(t_{k-1})(t_k - t_{k-1}).$$

Here the limit is taken over any division $0 = t_0 < t_1 < \dots < t_n = t$ with $\max_k(t_k - t_{k-1}) \to 0$. Show that the function

$$\eta(t) = \int_{t_0}^t \xi(s)ds, \quad t \geqslant t_0,$$

is differentiable in quadratic mean and that its derivative is $\eta'(t) = \xi(t)$.

Problem: Let the random function $\xi(t)$ be integrable in quadratic mean on the time interval T; show that it is integrable on any measurable set $\Delta \subseteq T$. More precisely, show that on the time interval T the random function $1_\Delta \xi(t)$ is integrable in quadratic mean, where 1_Δ is the indicator of the set Δ :

$$1_\Delta(t) = \begin{cases} 1, & t \in \Delta \\ 0, & t \notin \Delta. \end{cases}$$

Prove that the *integral*

(7.16) $\int_\Delta \xi(t)dt = \int_T 1_\Delta \cdot \xi(t)dt$

has the properties (7.10) - (7.13).

Section 8
Stochastic Measures and Integrals

The usual tools of mathematical analysis and of the theory of ordinary differential equations cannot be applied to random functions of the type of the Brownian motion process, which arise in several branches of probability theory and are important for application. The reason is that these functions turn out to be not differentiable. In the theory of random processes we apply the theory of stochastic analysis and stochastic differential equations, the basic element of which is the *stochastic integral*, which we shall deal with now.

Let T be a finite or infinite interval on the real line. On half-intervals of the form $\Delta = (s,t] \subseteq T$ let be given a function $\eta(\Delta)$ with values $\eta(\Delta) \in H$ in the Hilbert space H of random variables ξ, $M|\xi|^2 < \infty$, which has the following properties: for arbitrary disjoint Δ_1, Δ_2, the variables $\eta(\Delta_1)$, $\eta(\Delta_2)$ are *orthogonal*, i.e.

(8.1) $(\eta(\Delta_1),\ \eta(\Delta_2)) = 0.$

If $\Delta = \Delta_1 \cup \Delta_2$ is a halfinterval, consisting of two disjoint halfintervals Δ_1, Δ_2, we set

(8.2) $\eta(\Delta_1 \cup \Delta_2) = \eta(\Delta_1) + \eta(\Delta_2)$

and finally

(8.3) $\|\eta(\Delta)\|^2 = |\Delta|,$

where $|\Delta| = t - s$ for $\Delta = (s,t]$. The definition of the scalar product and the norm is the same as in (7.1) and (7.2) for the Hilbert space H. We extend the *additive function* $\eta(\Delta)$ to the ring of sets Δ, consisting of the unions of a finite number of disjoint halfintervals of the form $\Delta_k = (s_k,t_k]$, setting

(8.4) $\eta(\Delta) = \sum\limits_{k} \eta(\Delta_k)$

for an arbitrary union $\Delta = \cup_k \Delta_k$; from the orthogonality condition (8.1) follows that

$$\|\eta(\Delta)\|^2 = \sum\limits_{k,j} (\eta(\Delta_k), \eta(\Delta_j)) = \sum\limits_{k} \|\eta(\Delta_k)\|^2 = \sum\limits_{k} |\Delta_k|,$$

which may be expressed by the equation

(8.5) $\|\eta(\Delta)\|^2 = \mathbf{M}|\eta(\Delta)|^2 = \int_\Delta dt$

or, symbolically,

(8.6) $\mathbf{M}|\eta(dt)|^2 = dt.$

We shall call $\eta(\Delta)$, $\Delta \subseteq T$, a *stochastic additive function with orthogonal values*. We define the stochastic integral $\int_T \varphi(t)\eta(dt)$ for non-random functions $\varphi(t)$, satisfying the condition

(8.7) $\int_T |\varphi(t)|^2 dt < \infty.$

Let us begin with the piecewise constant function $\varphi(t)$, that takes only a finite number of values different from 0 on the disjoint half-intervals $\Delta_k \subseteq T$,

(8.8) $\varphi(t) = y_k, \quad t \in \Delta_k.$

For such a function, we define the stochastic integral by the equation

(8.9) $\int_T \varphi(t)\eta(dt) = \sum\limits_{k} y_k \cdot \eta(\Delta_k).$

Obviously, for arbitrary functions φ_1, φ_2 of the type described above and for the constants c_1, c_2, we can derive

(8.10) $\int_T (c_1\varphi_1(t) + c_2\varphi_2(t))\eta(dt) = c_1 \int_T \varphi_1(t)\eta(dt)$
$$+ c_2 \int_T \varphi_2(t)\eta(dt).$$

Applying the orthogonality condition (8.1) to the integral (8.9), we easily obtain the following equation:

(8.11) $\left\| \int_T \varphi(t)\eta(dt) \right\|^2 = \int_T |\varphi(t)|^2 dt$

and

(8.12) $\left[\int_T \varphi_1(t)\eta(dt), \int_T \varphi_2(t)\eta(dt) \right] = \int_T \varphi_1(t)\overline{\varphi_2(t)}\, dt$

for arbitrary φ_1, φ_2.

The left part of formula (8.11) may be written as

$$\left\| \sum_k y_k \cdot \eta(\Delta_k) \right\|^2 = \sum_{k,j} y_k \bar{y}_j (\eta(\Delta_k), \eta(\Delta_j))$$

$$= \sum_k |y_k|^2 \|\eta(\Delta_k)\|^2 = \sum_k |y_k|^2 |\Delta_k|,$$

where the last expression determines the integral on the right side of (8.11).

Now we take an arbitrary measurable function $\varphi(t)$ that satisfies the condition (8.7) and make use of the fact, that there exists a sequence of piecewise constant functions $\varphi_n(t)$ of the type (8.8), approximating the function $\varphi(t)$, such that

$$\int_T |\varphi(t) - \varphi_n(t)|^2 dt \to 0.$$

If we consider the sequence of corresponding integrals $\int_T \varphi_n(t)\eta(dt)$, then according to the general formula (8.11), we obtain

$$\left\| \int_T \varphi_n(t)\eta(dt) - \int_T \varphi_m(t)\eta(dt) \right\|^2$$

$$= \left\| \int_T (\varphi_n(t) - \varphi_m(t))\eta(dt) \right\|^2 = \int_T |\varphi_n(t) - \varphi_m(t)|^2 dt$$

$$\leqslant 2 \int_T (|\varphi_n(t) - \varphi(t)|^2 + |\varphi(t) - \varphi_m(t)|^2) dt \to 0.$$

for $n, m \to \infty$, i.e., this sequence turns out to be fundamental in the Hilbert space H. Hence, in H the limit

$$(8.13) \qquad \int_T \varphi(t)\eta(dt) = \lim_{n \to \infty} \int_T \varphi_n(t)\eta(dt),$$

exists, which defines the *stochastic integral* on the left. Obviously, this definition does not depend on the choice of the approximating sequence $\varphi_n(t)$ of the piecewise constant function $\varphi_n(t)$. (Verify this!)

Problem: Show that for continuous functions $\varphi(t)$, $t \geqslant 0$, on an arbitrary interval $T = [t_0, t]$, we can derive

$$\int_{t_0}^t \varphi(s)\eta(ds) = \lim_{n \to \infty} \sum_{k=1}^n \varphi(t_{k-1})\eta(\Delta_k),$$

where the limit is found from a sequence $t_0 < t_1 < \dots < t_n = t$ on the half-interval $\Delta_k = (t_{k-1}, t_k]$ with $\max_k(t_k - t_{k-1}) \to 0$.

Problem: Show that the formulas (8.10) - (8.12) can be extended to arbitrary measurable functions $\varphi(t)$, satisfying condition (8.7).

Problem: For an arbitrary bounded measurable set $\Delta \subseteq T$ we define

(8.14) $n(\Delta) = \int_T 1_\Delta \cdot n(dt).$

The right-hand side is the stochastic integral of the indicator $\varphi(t) = 1_\Delta$ of the set Δ. Show that formula (8.14) defines an additive stochastic function with orthogonal values, which satisfies the conditions (8.1) - (8.3) for all $\Delta \subseteq T$ and, moreover, has the property of σ-additivity:

(8.15) $n(\Delta) = \sum_k n(\Delta_k) = \lim_{n \to \infty} \sum_{k=1}^{n} n(\Delta_k)$

for a countable number of disjoint sets Δ_k, $\cup_k \Delta_k = \Delta$ (such an additive function is called *stochastic measure with orthogonal values*).

Problem: Let $\xi(t)$, $t \geqslant 0$, be a Poisson process with parameter $\lambda = 1$:

$$\mathbf{M}(\xi(t) - \xi(s)) = t - s.$$

For the half-interval $\Delta = (s,t]$ we set

$$n(\Delta) = \xi(t) - \xi(s) - (t - s).$$

Prove that the function $n(\Delta) \in \mathbf{H}$ has the properties (8.1) - (8.3). Show that for continuous functions $\varphi(t)$, $t \geqslant 0$,

$$\int_0^t \varphi(s)n(ds) = \sum_{0 < \tau_k < t} \varphi(\tau_k) - \int_0^t \varphi(s)ds,$$

with probability 1, where τ_k are the random times of jumps of the Poisson process. ($\tau_k = \Delta_0 + \dots + \Delta_{k-1}$, cf. Fig. 1 on page 4).

From now on, we shall consider stochastic measures with mean value

(8.16) $\mathbf{M}n(\Delta) = (n(\Delta), 1) = 0.$

Problem: Show that under condition (8.16) the equation

(8.17) $\mathbf{M} \int_T \varphi(t)n(dt) = \left[\int_T \varphi(t)n(dt), 1 \right] = 0.$

is true.

Hint: Apply equation (8.17) for piecewise constant functions.

Section 9
The Stochastic Ito Integral and Stochastic Differentials

Suppose that a σ-algebra of events B^t is given; we interpret it as the total number of events up to time t. Then, accordingly,

(9.1) $\qquad B^s \subseteq B^t, \quad s \leqslant t.$

We generalize the idea of the stochastic integral to random functions $\varphi(t)$, the values of which at each time t are the random variables

$$\varphi(t) = \varphi(\omega, t), \quad \omega \in \Omega,$$

on the space of elementary events Ω; these functions are measurable with respect to the corresponding σ-algebra of events B^t. We shall call such random functions *non-anticipating*. We suppose that the stochastic measure $\eta(\Delta)$ with the properties described in (8.1) - (8.5) has the mean value

(9.2) $\qquad \mathbf{M}\eta(\Delta) = 0,$

where for each $\Delta = (s, t]$ the random variable $\eta(\Delta)$ is measurable with respect to the σ-algebra of events B^t and does not depend on the σ-algebra of events B^s up to time s.

As before, we shall deal with random variables ξ, $\mathbf{M}|\xi|^2 < \infty$, in the Hilbert space $\mathbf{H}$ with scalar product (7.1) and we set $\xi_1 = \xi_2$, if the variables ξ_1, ξ_2 are equal with probability 1.

We begin the definition of the stochastic integral $\int_t \varphi(t)\eta(dt)$ in the time interval T be considering piecewise constant random functions $\varphi(t)$, $t \in T$, that take only a finite number of values different from zero on the half-intervals of the form $\Delta_k = (s_k, t_k]$ say,

(9.3) $\qquad \varphi(t) = \xi_k, \quad t \in \Delta_k,$

where each of the random variables $\xi_k \in \mathbf{H}$ is measurable with respect to the corresponding σ algebra B^{s_k}.

Problem: Let the σ-algebra B^t, $t \geq t_0$, be *right continuous*; more precisely, for all s

(9.4) $\bigcap_{t>s} B^t = B^s$.

Show that for each non-anticipating piecewise constant function of

the form (9.3), the values ξ_k are measurable with respect to B^{s_k}.

From now on, we shall suppose that condition (9.4) is satisfied. For non-anticipating piecewise constant functions $\varphi(t)$ of the form (9.3), we define the stochastic integral by the equation

(9.5) $\int_T \varphi(t)\eta(dt) = \sum_k \xi_k \cdot \eta(\Delta_k)$.

Under these conditions, the variables ξ_k and $\eta(\Delta_k)$ are independent and therefore

$$\mathbf{M}|\xi_k \cdot \eta(\Delta_k)|^2 = \mathbf{M}|\xi_k|^2 \cdot \mathbf{M}|\eta(\Delta_k)|^2 = \|\xi_k\|^2|\Delta_k|.$$

Obviously, the stochastic integral (9.5) is an element of the Hilbert space **H**. We show that under condition (9.2)

(9.6) $\mathbf{M} \int_T \varphi(t)\eta(dt) = \left[\int_T \varphi(t)\eta(dt), 1\right] = 0.$

Indeed, the variables $\eta(\Delta_k)$ do not depend on the corresponding ξ_k, and

$$\mathbf{M} \sum_k \xi_k \eta(\Delta_k) = \sum_k \mathbf{M}\xi_k \cdot \mathbf{M}\eta(\Delta_k) = 0.$$

It is easy to prove that

(9.7) $\left\|\int_T \varphi(t)\eta(dt)\right\|^2 = \int_T \|\varphi(t)\|^2 dt.$

In fact, for $k > j$, the variable $\eta(\Delta_k)$ does not depend on any variable ξ_j, $\eta(\Delta_j)$, ξ_k and we have

$$\mathbf{M}[\xi_j \eta(\Delta_j)\xi_k \eta(\Delta_k)] = \mathbf{M}[\xi_j \eta(\Delta_j)\xi_k] \cdot \mathbf{M}\eta(\Delta_k) = 0.$$

(As for the existence of these mathematical expectations, we remind you that the variables $\xi = \xi_j \eta(\Delta_j)$, $\xi_k \eta(\Delta_k)$, ξ_k, $\eta(\Delta_k)$ have a finite second moment $\mathbf{M}|\xi|^2 < \infty$.) And, for $k \neq j$ the variables $\xi_j \eta(\Delta_j)$, $\xi_k \eta(\Delta_k) \in \mathbf{H}$, are orthogonal. Therefore

$$\left\|\sum_k \xi_k \eta(\Delta_k)\right\|^2 = \sum_k \|\xi_k \eta(\Delta_k)\|^2 = \sum_k \|\xi_k\|^2|\Delta_k|$$

and, hence, formula (9.7) is true.

Obviously, a linear combination of non-anticipating piecewise constant functions $\varphi(t)$ is a function of the same type. This can easily be seen from the representation (9.3) for $\varphi = \varphi_1$, $\varphi = \varphi_2$ with the same half-intervals Δ_k for φ_1, φ_2. Clearly, we have

(9.8)
$$\int_T (c_1\varphi_1(t) + c_2\varphi_2(t))\eta(dt)$$
$$= c_1 \int_T \varphi_1(t)\eta(dt) + c_2 \int_T \varphi_2(t)\eta(dt).$$

Let us consider a non-anticipating random function $\varphi(t)$, $\int_T \|\varphi(t)\|^2 dt < \infty$, for which exists a sequence of non-anticipating piecewise constant functions $\varphi_n(t)$ that converges to $\varphi(t)$, such that

(9.9)
$$\lim_{n\to\infty} \int_T \|\varphi(t) - \varphi_n(t)\|^2 dt = 0.$$

The corresponding sequence of stochastic integrals $\int_T \varphi_n(t)\eta(dt)$ is fundamental, since, according to the general formula (9.6),

$$\left\| \int_T \varphi_n(t)\eta(dt) - \int_T \varphi_m(t)\eta(dt) \right\|^2$$

$$= \left\| \int_T (\varphi_n(t) - \varphi_m(t))\eta(dt) \right\|^2 = \int_T \|\varphi_n(t) - \varphi_m(t)\|^2 dt$$

$$\leqslant 2\int_T (\|\varphi_n(t) - \varphi(t)\|^2 + \|\varphi(t) - \varphi_m(t)\|^2)dt \to 0$$

for $n,m \to \infty$. Hence, in the Hilbert space H the limit in quadratic mean exists:

(9.10)
$$\int_T \varphi(t)\eta(dt) = \lim_{n\to\infty} \int_T \varphi_n(t)\eta(dt).$$

Obviously, this limit does not depend on the choice of the approximating sequence of non-anticipating, piecewise constant functions $\varphi_n(t)$ (Verify this!). The limit in (9.10) defines the *stochastic integral*, which is an extension of the stochastic integral (8.6) to *non-anticipating* random functions $\varphi(t)$, satisfying condition (9.9).

Problem: Show that the formulas (9.6) - (9.8) can be extended to arbitrary non-anticipating random functions $\varphi(t)$, satisfying condition (9.9).

Henceforth we shall assume that our stochastic measure $\eta(\Delta)$ is given by the standard Wiener process $\eta(t)$, $t \geqslant t_0$, via

(9.11)
$$\eta(\Delta) = \eta(t) - \eta(s), \quad \Delta = (s,t].$$

In this case, the stochastic integral (9.10) is usually called stochastic Ito integral, for which we choose the notation $d\eta(t)$ instead of $\eta(dt)$, because it is more adequate to definition (9.11).

Problem: Show that for non-anticipating continuous (in quadratic mean) functions $\varphi(t)$, $t \geqslant 0$, the stochastic Ito integral

(9.12) $\int_{t_0}^{t} \varphi(s) d\eta(s) = \lim\limits_{n \to \infty} \sum\limits_{k=1}^{n} \varphi(t_{k-1})[\eta(t_k) - \eta(t_{k-1})],$

exists, where the limit is taken over a division $t_0 < t_1 < ... < t_n = t$ with $\max_k(t_k - t_{k-1}) \to 0$. Show that the random function

$$\xi(t) = \int_{t_0}^{t} \varphi(s) d\eta(s), \quad t \geqslant t_0.$$

has the following property: for $\varphi(t) \neq 0$ and small $h \to 0$,

$$\xi(t + h) - \xi(t) = \varphi(t)[\eta(t + h) - \eta(t)] + o(h^{1/2}),$$

where the remainder term $\|o(h^{1/2})\|$ is negligeable compared to

$$\|\varphi(t)[\eta(t + h) - \eta(t)]\| = \|\varphi(t)\| \cdot h^{1/2}.$$

To show the characteristic feature of the stochastic Ito integral, we shall compute it for the standard Wiener process $\varphi(t) = \eta(t)$, $t \geqslant 0$, applying formula (9.12):

$$\int_{0}^{t} \eta(s) d\eta(s) = \lim\limits_{n \to \infty} \sum\limits_{k=1}^{n} \eta(t_{k-1})[\eta(t_k) - \eta(t_{k-1})].$$

Using equation

$$\eta(t_{k-1})[\eta(t_k) - \eta(t_{k-1})]$$
$$= \frac{1}{2}[\eta(t_k)^2 - \eta(t_{k-1})^2] - \frac{1}{2}[\eta(t_k) - \eta(t_{k-1})]^2,$$

we obtain

$$\sum\limits_{k=1}^{n} \eta(t_{k-1})[\eta(t_k) - \eta(t_{k-1})]$$
$$= \frac{1}{2}[\eta(t)^2 - \eta(t_0)^2] - \frac{1}{2}\sum\limits_{k=1}^{n}[\eta(t_k) - \eta(t_{k-1})]^2,$$

where $t_0 = 0$, $\eta(t_0) = 0$. Recall that for the standard Wiener process $\eta(t)$, $t \geqslant 0$, the limit in quadratic mean

(9.13) $\lim\limits_{n \to \infty} \sum\limits_{k=1}^{n} [\eta(t_k) - \eta(t_{k-1})]^2 = t$

exists, where $\max_k|t_k - t_{k-1}| \to 0$. (Cf. relation (5.19) and its demonstration.) Therefore

(9.14) $\int_{0}^{t} \eta(s) d\eta(s) = \frac{1}{2}\eta(t)^2 - \frac{1}{2}t.$

The random function $\xi(t)$, $t \geqslant t_0$, is said to have the *stochastic differential*

(9.15) $d\xi(t) = \alpha(t)dt + \beta(t)d\eta(t),$

if

(9.16) $\xi(t) = \xi(t_0) + \int_{t_0}^{t} \alpha(s)ds + \int_{t_0}^{t} \beta(s)d\eta(s), \quad t \geqslant t_0,$

where $\alpha(t)$, $\beta(t)$ are non-anticipating random functions (and so the stochastic integral is meaningful).

Problem: Show that a random process having the stochastic differential (9.15) is continuous in quadratic mean.

Hint: Use the relation

$$\int_{s}^{s+h} \|\alpha(t)\|dt \to 0, \quad \int_{s}^{s+h} \|\beta(t)\|^2 dt \to 0,$$

which holds for integrable functions

$$\|\alpha(t)\| = (\mathbf{M}|\alpha(t)|^2)^{1/2}, \quad \|\beta(t)\|^2 = \mathbf{M}|\beta(t)|^2$$

for $h \to 0$.

Problem: Suppose that the random function $\xi(t)$, $t \geqslant s$, has a stochastic differential (9.15). Show that for an arbitrary event $A \in \mathcal{B}^s$, the random function $\xi(t)1_A$, $t \geqslant s$, has the stochastic differential

(9.17) $d[\xi(t) \cdot 1_A] = [\alpha(t) \cdot 1_A]dt + [\beta(t) \cdot 1_A]d\eta(t),$

where the random variable $1_A(\omega)$, $\omega \in \Omega$, is the indicator of the event A:

$$1_A(\omega) = \begin{cases} 1, & \omega \in A, \\ 0, & \omega \notin A. \end{cases}$$

Hint: Show that the constant factor $1_A(\omega)$ may be taken outside the integral sign:

$$\int_{s}^{t} [1_A \cdot \alpha(u)]du = 1_A \cdot \int_{s}^{t} \alpha(u)du,$$

$$\int_{s}^{t} [1_A \cdot \beta(u)]d\eta(u) = 1_A \cdot \int_{s}^{t} \beta(u)d\eta(u).$$

Problem: Show that the random function $\xi(t) = \eta(t)^2$, where $\eta(t)$, $t \geqslant 0$, is the standard Wiener process, has the stochastic differential

$$d\xi(t) = dt + 2\eta(t)d\eta(t).$$

Hint: Apply formula (9.14).

Let us calculate the stochastic differential of a random process of the form

$$(9.18) \qquad \xi(t) = \int_{t_0}^t c(t,s)d\eta(s), \quad t \geqslant t_0,$$

where $c(t,s)$ is a nonrandom function of the variables $t \geqslant s \geqslant t_0$ with the derivative $(d/dt)c(t,s)$ that is continuous relative to all variables. If we interchange the order of integration, we obtain the equation

$$\int_{t_0}^t \left[\int_{t_0}^u \frac{d}{du} c(u,s)d\eta(s) \right] du = \int_{t_0}^t \left[\int_s^t \frac{d}{du} c(u,s)du \right] d\eta(s)$$

$$= \int_{t_0}^t [c(t,s) - c(s,s)]d\eta(s) = \xi(t) - \int_{t_0}^t c(s,s)d\eta(s).$$

Therefore,

$$(9.19) \qquad d\xi(t) = \left[\int_{t_0}^t \frac{d}{dt}c(t,s)d\eta(s) \right] dt + c(t,t)d\eta(t).$$

In order to justify the change of the order of integration that we applied above, we prove a general *formula of interchange of the integration order*. Let $\varphi(u,s)$ be a nonrandom measurable function of variables $t_0 \leqslant u, s \leqslant t$, for which the stochastic integrals below exist:

$$(9.20) \qquad \eta_1 = \int_{t_0}^t \left[\int_{t_0}^t \varphi(u,s)du \right] d\eta(s), \quad \eta_2 = \int_{t_0}^t \left[\int_{t_0}^t \varphi(u,s)d\eta(s) \right] du.$$

(Suppose these integrals exist for the function

$$\varphi(u,s) = \begin{cases} \dfrac{d}{du} c(u,s), & t_0 \leqslant s \leqslant u, \\ 0 & , \quad u < s \leqslant t, \end{cases}$$

which we deal with if we calculate the stochastic differential (9.19).)

According to the definition of the stochastic integral, the variables (9.20) belong to the closure (in the Hilbert space H) of the space of all possible linear combinations of variables of the type $\eta = \eta(t_2) - \eta(t_1)$, where $t_0 \leqslant t_1 < t_2 \leqslant t$. Therefore, the equation $\eta_1 = \eta_2$ holds, if we can prove that equation $(\eta_1,\eta) = (\eta_2,\eta)$ holds for their scalar products. We have

$$(\eta_1,\eta) = \int_{t_1}^{t_2} \left[\int_{t_0}^t \varphi(u,s)du \right] ds,$$

$$(\eta_2,\eta) = \int_{t_0}^t \left[\int_{t_0}^t \varphi(u,s)d\eta(s),\eta \right] du = \int_{t_0}^t \left[\int_{t_1}^{t_2} \varphi(u,s)ds \right] du.$$

Since these are ordinary double integrals that fulfill the equation

$$\int_{t_0}^{t} \left[\int_{t_1}^{t_2} \varphi(u,s)ds \right] du = \int_{t_1}^{t_2} \left[\int_{t_0}^{t} \varphi(u,s)du \right] ds$$

we can derive

(9.21) $$\int_{t_0}^{t} \left[\int_{t_0}^{t} \varphi(u,s)du \right] d\eta(s) = \int_{t_0}^{t} \left[\int_{t_0}^{t} \varphi(u,s)d\eta(s) \right] du,$$

which was to be demonstrated.   □

Section 10
Stochastic Differential Equations

In this paragraph we shall consider (real) random processes $\xi(t)$, $t \geqslant t_0$, characterized by the stochastic differential

$$d\xi(t) = \alpha(t)dt + \beta(t)d\eta(t),$$

$$\alpha(t) = a(t,\xi(t)), \quad \beta(t) = b(t,\xi(t)).$$

where $a(t,x)$, $b(t,x)$ are non-random functions of the parameters $t \geqslant t_0$ and $-\infty < x < \infty$.

Given the functions $a(t,x)$, $b(t,x)$, we want to know whether a random process of this type with stochastic differential

$$(10.1) \qquad d\xi(t) = a(t, \xi(t))dt + b(t, \xi(t))d\eta(t)$$

exists, or whether the solution exists of the stochastic integral equation

$$(10.2) \qquad \xi(t) = \xi(t_0) + \int_{t_0}^t a(s,\xi(s))ds + \int_{t_0}^t b(s,\xi(s))d\eta(s), \quad t \geqslant t_0.$$

If we write this last equation in symbolic form with differentials, we obtain the stochastic differential equation (10.1).

The functions $\alpha(t) = a(t,\xi(t))$, $\beta(t) = b(t,\xi(t))$ must satisfy the conditions under which we defined the stochastic integrals like in (10.2). Recall that the random function

$$\xi(t) = \xi(t_0) + \int_{t_0}^t \alpha(s)ds + \int_{t_0}^t \beta(s)d\eta(s), \quad t \geqslant t_0,$$

is continuous in quadratic mean, if both $\alpha(t)$, $\beta(t)$ satisfy these conditions. Having this in mind, we suppose that our coefficients $a(t,x)$, $b(t,x)$ have the following properties:

$$|a(t,x) - a(s,x)| \leqslant C(1 + |x|) |a(t) - a(s)|,$$
(10.3)
$$|b(t,x) - b(s,x)| \leqslant C(1 + |x|) |b(t) - b(s)|,$$

where $a(t)$, $b(t)$ are continuous functions. Furthermore, let us assume that

$$|a(t,x) - a(t,y)| \leqslant C|x - y|,$$
(10.4)
$$|b(t,x) - b(t,y)| \leqslant C|x - y|$$

for all x, y uniformly with respect to t on each finite interval $t_0 \leqslant t \leqslant T$ with a corresponding constant C.

Problem: Show that under the conditions (10.3), (10.4) for any random function $\xi(t)$, $t \geqslant t_0$, which is continuous in quadratic mean, the random functions

$$\alpha(t) = a(t, \xi(t)), \quad \beta(t) = b(t, \xi(t))$$

are continuous in quadratic mean.

Recall that the definition of the stochastic Ito integral in (10.2) requires the assumption that the random function $\beta(t) = b(t,\xi(t))$ is non-anticipating. Hence, when talking of a random function with stochastic differential (10.1), we mean a non-anticipating random function $\xi(t)$, $t \geqslant t_0$, with the random variables $\xi(t)$ being measurable with regard to the σ-algebra of events B^t up to the corresponding time t for every t. In particular, we suppose that the random variable $\xi(t_0)$ at the initial time t_0 is non-anticipating. The random processes of the type (10.1) with non-degenerated coefficient $b(t,x) \neq 0$ have the following character locally. Imagine that $\xi(t)$, $t \geqslant t_0$, describes the motion of some particle; then for $\xi(s) = x$, the mean increment of the trajectory of the particle after a short time interval h is

$$a(s,x)h + o(h)$$

-- approximately the same as for a particle moving at a uniform speed $a(s,x)$, depending on the initial position $\xi(s) = x$. This linear motion is disturbed by a random fluctuation

$$b(s,x)[\eta(s + h) - \eta(s)] + o(h^{1/2})$$

-- approximately the same as for the Brownian motion with diffusion coefficient

$$\sigma^2 = b(s,x)^2.$$

Theorem. *For arbitrary initial variables $\xi(t_0)$, the solution $\xi(t)$, $t \geqslant t_0$, of the stochastic differential equation (10.1) exists and is unique.*

Proof: We use the method of sequential approximation in our Hilbert space **H**, setting

$$\xi_0(t) = \xi(t_0),$$

$$\xi_1(t) = \xi(t_0) + \int_{t_0}^t a(s,\xi_0(s))ds + \int_{t_0}^t b(s,\xi_0(s))d\eta(s),$$

(10.5) $\cdot$ $\cdot$ $\cdot$ $\cdot$ $\cdot$ $\cdot$ $\cdot$ $\cdot$ $\cdot$ $\cdot$ $\cdot$ $\cdot$ $\cdot$ $\cdot$ $\cdot$ $\cdot$ $\cdot$ $\cdot$ $\cdot$ $\cdot$ $\cdot$

$$\xi_n(t) = \xi(t_0) + \int_{t_0}^t a(s,\xi_{n-1}(s))ds + \int_{t_0}^t b(s,\xi_{n-1}(s))d\eta(s),$$

$\cdot$ $\cdot$ $\cdot$ $\cdot$ $\cdot$ $\cdot$ $\cdot$ $\cdot$ $\cdot$ $\cdot$ $\cdot$ $\cdot$ $\cdot$ $\cdot$ $\cdot$ $\cdot$ $\cdot$ $\cdot$ $\cdot$ $\cdot$ $\cdot$

where, under condition (10.3), (10.4) all $\xi_n(t)$ and $a(t,\xi_n(t))$, $b(t,\xi_n(t))$ are non-anticipating functions depending on t, which are continuous in quadratic mean (Proof!). With condition (10.4), we get

$$\|\xi_{n+1}(t) - \xi_n(t)\|$$

$$\leqslant \left\| \int_{t_0}^t [a(s,\xi_n(s)) - a(s,\xi_{n-1}(s))]ds \right\|$$

$$+ \left\| \int_{t_0}^t [b(s,\xi_n(s)) - b(s,\xi_{n-1}(s))]d\eta(s) \right\|$$

$$\leqslant \int_{t_0}^t \|a(s,\xi_n(s)) - a(s,\xi_{n-1}(s))\|ds$$

$$+ \left[\int_{t_0}^t \|b(s,\xi_n(s)) - b(s,\xi_{n-1}(s))\|^2 ds \right]^{1/2}$$

$$\leqslant C \max_{t_0 \leqslant s \leqslant t} \|\xi_n(s) - \xi_{n-1}(s)\|(t-t_0) + C \max_{t_0 \leqslant s \leqslant t} \|\xi_n(s) - \xi_{n-1}(s)\|(t-t_0)^{1/2}.$$

The constant C in condition (10.4) may depend on the corresponding interval $t_0 \leqslant t \leqslant T$. Let $t_0 < t_1 < \ldots < t_m = T$ and set $l = \max_k |t_k - t_{k-1}|$. Assuming l to be so small that $C(l + \sqrt{l}) = r < 1$, we obtain the following estimation:

$$\max_{t_0 \leqslant t \leqslant t_1} \|\xi_{n+1}(t) - \xi_n(t)\| \leqslant r \max_{t_0 \leqslant t \leqslant t_1} \|\xi_n(t) - \xi_{n-1}(t)\| \leqslant \cdots \leqslant$$

$$\leqslant r^n \max_{t_0 \leqslant t \leqslant t_1} \|\xi_1(t) - \xi_0(t)\| = C_0 r^n.$$

As a final result follows for arbitrary $m > n$

$$\max_{t_0 \leqslant t \leqslant t_1} \|\xi_m(t) - \xi_n(t)\| \leqslant C_0 \sum_{k=n}^{m-1} r^k \leqslant C_1 r^n \to 0, \quad m,n \to \infty .$$

In particular, it is obvious that the sequence of variables $\xi_n(t)$, $n = 0,1, ...$, is fundamental in $\mathbf{H}$, and that for each t, $t_0 \leqslant t \leqslant t_1$, the limit in quadratic mean exists:

(10.6) $\qquad \xi(t) = \lim_{n \to \infty} \xi_n(t).$

It is evident that

$$\max_{t_0 \leqslant t \leqslant t_1} \|\xi(t) - \xi_n(t)\| \leqslant C_1 r^n .$$

Moreover with condition (10.4)

$$\max_{t_0 \leqslant t \leqslant t_1} \|a(t,\xi(t)) - a(t,\xi_n(t))\| \leqslant C_2 r^n ,$$

$$\max_{t_0 \leqslant t \leqslant t_1} \|b(t,\xi(t)) - b(t,\xi_n(t))\| \leqslant C_2 r^n .$$

As the random functions $\xi_n(t)$ are continuous in quadratic mean, their uniform limit in quadratic mean $\xi(t)$, $t_0 \leqslant t \leqslant t_1$, has the same property. Obviously, the limit function $\xi(t)$ satisfies the integral equation (10.2):

$$\xi(t) = \lim_{n \to \infty} \xi_n(t)$$

$$= \xi(t_0) + \lim_{n \to \infty} \int_{t_0}^{t} a(s,\xi_{n-1}(s))ds$$

$$+ \lim_{n \to \infty} \int_{t_0}^{t} b(s,\xi_{n-1}(s))d\eta(s)$$

$$= \xi(t_0) + \int_{t_0}^{t} a(s,\xi(s))ds + \int_{t_0}^{t} b(s,\xi(s))d\eta(s).$$

This function $\xi(t)$, $t_0 \leqslant t \leqslant t_1$, is the unique solution of the integral equation (10.2). If we had another solution $\tilde{\xi}(t)$, $t_0 \leqslant t \leqslant t_1$, then

$$\|\xi(t) - \tilde{\xi}(t)\| \leqslant \left\| \int_{t_0}^{t} [a(s,\xi(s)) - a(s,\tilde{\xi}(s))]ds \right\|$$

$$+ \left\| \int_{t_0}^{t} [b(s,\xi(s)) - b(s,\tilde{\xi}(s))]d\eta(s) \right\|$$

$$\leqslant r \cdot \max_{t_0 \leqslant s \leqslant t_1} \|\xi(s) - \tilde{\xi}(s)\|, \quad r < 1,$$

which is only possible, if

$$\max_{t_0 \leqslant s \leqslant t_1} \|\xi(s) - \tilde{\xi}(s)\| = 0.$$

It is obvious that this method of sequential approximation enables us to find the solution with given initial value $\xi(t_1)$ on the subsequent interval $t_1 \leqslant t \leqslant t_2$ and so on; and as the result, we obtain the solution $\xi(t)$, $t_0 \leqslant t \leqslant T$, which is the same for each finite interval $t_0 \leqslant t \leqslant T$. The theorem is demonstrated. $\qquad$ $\square$

Problem: Let $\xi(t)$, $\check{\xi}(t)$ be solutions of equation (10.1) with initial values $\xi(t_0)$, $\check{\xi}(t_0)$. Show that the estimation

$$(10.7) \qquad \| \xi(t) - \check{\xi}(t) \| \leqslant C \cdot \| \xi(t_0) - \check{\xi}(t_0) \|$$

holds on any finite interval $t_0 \leqslant t \leqslant t_1$, where the constant C depends on t_1.

Hint: Estimate

$$\| \xi_n(t) - \check{\xi}_n(t) \|, \quad n = 0, 1, \dots .$$

Section 11
Diffusion Processes
Kolmogorov's Differential Equations

Suppose that the probability densities $p(s,x,t,y)$, $-\infty < y < \infty$, depend on the parameters $t > s > t_0$, $-\infty < x < \infty$ in such a way, that the *Kolmogorov-Chapman equation* holds:

$$(11.1) \qquad p(s,x,t,y) = \int_{-\infty}^{\infty} p(s,x,u,z)p(u,z,t,y)dz, \quad s < u < t.$$

Let us consider the random process $\xi(t)$, $t \geqslant t_0$ with initial value $\xi(t_0) = x_0$, where the variables $\xi(t_1)$, ..., $\xi(t_n)$ at times $t_0 < t_1 < ... < t_n$ are distributed in $\mathbb{R}^n$ with the corresponding density

$$(11.2) \qquad p_{t_1,...,t_n}(x_1, ..., x_n) = p(t_0,x_0,t_1,x_1) \cdots p(t_{n-1},x_{n-1},t_n,x_n),$$

$$(x_1, ..., x_n) \in \mathbb{R}^n.$$

The conditional distribution of the variables $\xi(t)$, $t \geqslant s$, under the condition

$$\xi(s_1) = x_1, ..., \xi(s_m) = x_m, \quad \xi(s) = x$$

for any $s_1 < ... < s_m < s$ is, independently of x_1, ..., x_m,

$$(11.3) \qquad P\{\xi(t) \in B \mid \xi(s) = x\} = \int_B p(s,x,t,y)dy, \quad B \subseteq \mathbb{R}^1$$

(Proof!) This is a so-called Markov process with transition density $p(s,x,t,y)$. If the process is in the state $\xi(s) = x$ at time s, then it will be in the state $\xi(t) \in B$ at time t with the probability given by (11.3). We assume that

$$(11.4) \qquad \int_{|y-x|>\epsilon} p(s, x, s + h, y)dy = o(h)$$

for $h \to 0$ for any fixed $\epsilon \to 0$; we also assume that

(11.5) $\int_{|y-x|\leqslant\epsilon} (y - x)p(s, x, s + h, y)dy = a(s, x)\cdot h + o(h)$

(11.6) $\int_{|y-x|\leqslant\epsilon} (y - x)^2 p(s, x, s + h, y)dy = b(s, x)^2 \cdot h + o(h),$

where $o(h)/h \to 0$ for $h \to 0$, uniformly on each finite interval $t_0 < s < t_1$. A random process $\xi(t)$, $t \geqslant t_0$, which satisfies the above mentioned properties, is usually called *diffusion process*, and the coefficients $a(s,x)$ and $b(s,x)^2$ appearing in (11.5) and (11.6) are called *drift-* and *diffusion coefficient* respectively.

An example of a process of this type with parameters $a(t,x) = 0$, $b(t,x)^2 = \sigma^2$ is the Brownian motion process, which we already considered. Recall the diffusion equation that we used there (cf. Sec. 5; verify, that the equations (11.4) - (11.6) hold for the transition density (5.6)).

Theorem. *Suppose that the transitions density $p(s,x,t,y)$ has the derivatives $\partial p/\partial s$, $\partial p/\partial x$ and $\partial^2 p/\partial x^2$, which are continuous with respect to x uniformly for all y on each finite interval $y_0 \leqslant y \leqslant y_1$. Then the probability density satisfies the diffusion equation*

(11.7) $-\dfrac{\partial p}{\partial s} = a(s,x)\dfrac{\partial p}{\partial x} + \dfrac{1}{2} b(s,x)^2 \dfrac{\partial^2 p}{\partial x^2}.$

Proof: We take an arbitrary continuous function $\varphi(x)$ equal to zero outside some finite interval, and we set

(11.8) $\varphi(s,x) = \displaystyle\int_{-\infty}^{\infty} \varphi(y)p(s,x,t,y)dy.$

It follows from the Kolmogorov-Chapman equation that for any $t_0 < s < u < t$

$$\varphi(s) = \int_{-\infty}^{\infty}\varphi(y) \int_{-\infty}^{\infty} p(s,x,u,z)p(u,z,t,y)dz\ dy$$

$$= \int_{-\infty}^{\infty}\varphi(u,z)p(s,x,u,z)dz.$$

Obviously, the function $\varphi(s,x)$ has continuous derivatives $\partial\varphi/\partial s$, $\partial\varphi/\partial x$, $\partial^2\varphi/\partial x^2$. We develop the function $\varphi(u,z)$ according to the Taylor formula in a neighbourhood of the point x (for fixed u):

$$\varphi(u,z) - \varphi(u,x)$$

$$= \dfrac{\partial\varphi(u,x)}{\partial x}(z - x) + \dfrac{1}{2}\left[\dfrac{\partial^2\varphi(u,x)}{\partial x^2} + O(\delta_\epsilon)\right](z - x)^2,$$

where

$$\delta_\epsilon = \sup_{|z-x|\leqslant\epsilon} \left|\dfrac{\partial^2\varphi(u,z)}{\partial x^2} - \dfrac{\partial^2\varphi(u,x)}{\partial x^2}\right| \to 0$$

for $\epsilon \to 0$. From the relations (11.4) - (11.6) we get

$$\varphi(s,x) \; - \; \varphi(u,x) \; = \; \int_{-\infty}^{\infty} [\varphi(u,z) \; - \; \varphi(u,x)] p(s,x,u,z) dz$$

$$= \int_{|z-x| \leqslant \epsilon} [\varphi(u,z) \; - \; \varphi(u,x)] p(s,x,u,z) dz \; + \; o(u \; - \; s)$$

$$= \frac{\partial \varphi(u,x)}{\partial x} \int_{|z-x| \leqslant \epsilon} (z \; - \; x) p(s,x,u,z) dz$$

$$+ \; \frac{1}{2} \left[\frac{\partial^2 \varphi(u,x)}{\partial x^2} \; + \; O(\delta_\epsilon) \right] \int_{|z-x| \leqslant \epsilon} (z \; - \; x)^2 p(s,x,u,z) dz$$

$$+ \; o(u \; - \; s)$$

$$= \left\{ a(s,x) \; \frac{\partial \varphi(u,x)}{\partial x} \; + \; \frac{1}{2} b(s,x)^2 \left[\frac{\partial^2 \varphi(u,x)}{\partial x^2} \right. \right.$$

$$\left. \left. + \; O(\delta_\epsilon) \right] \right\} (u \; - \; s) \; + \; o(u \; - \; s)$$

where $O(\delta_\epsilon) \to 0$ for $\epsilon \to 0$. It can be concluded that

$$\lim_{u \to s} \frac{\varphi(s,x) \; - \; \varphi(u,x)}{u \; - \; s} \; = \; a(s,x) \; \frac{\partial \varphi(s,x)}{\partial x} \; + \; \frac{1}{2} \; b(s,x)^2 \; \frac{\partial^2 \varphi(s,x)}{\partial x^2}$$

and, hence,

$$- \; \frac{\partial \varphi}{\partial s} = a(s,x) \; \frac{\partial \varphi}{\partial x} \; + \; \frac{1}{2} b(s,x)^2 \; \frac{\partial^2 \varphi}{\partial x^2} \; .$$

Recalling the definition of the function $\varphi(s,x)$ (cf. (11.8)), we can rewrite the equation in the following way:

$$\int_{-\infty}^{\infty} \varphi(y) \left[\frac{\partial p}{\partial s} \; + \; a(s,x) \; \frac{\partial p}{\partial x} \; + \; \frac{1}{2} \; b(s,x)^2 \; \frac{\partial^2 p}{\partial x^2} \right] dy \; = \; 0,$$

where $\varphi(y)$ is an arbitrary continuous function equal to zero outside some finite interval, and, hence, the equation

$$\frac{\partial p}{\partial s} + \; a(s,x) \frac{\partial p}{\partial x} \; + \; \frac{1}{2} b(s,x)^2 \; \frac{\partial^2 p}{\partial x^2} \; = \; 0$$

has to be satisfied. The theorem is proved. $\square$

Theorem. *Suppose that the derivatives*

$$\frac{\partial}{\partial t} \; p(s,x,t,y), \qquad \frac{\partial}{\partial y} [a(t,y) p(s,x,t,y)], \qquad \frac{\partial^2}{\partial y^2} [b(t,y)^2 p(s,x,t,y)].$$

exist and are continuous. Then the transition density $p(s,x,t,y)$ satisfies the differential equation

$$(11.9) \qquad \frac{\partial p}{\partial t} \; = \; - \; \frac{\partial}{\partial y} \; [a(t,y) p(s,x,t,y)] + \frac{1}{2} \frac{\partial^2}{\partial y^2} \; [b(t,y)^2 p(s,x,t,y)].$$

Proof: Replicating exactly the proof of the previous theorem, we can derive that for any twice continuously differentiable function $\varphi(x)$, equal to zero outside some finite interval, the limit

$$\lim_{h \to 0} \frac{1}{h} \left[\int_{-\infty}^{\infty} \varphi(y) p(t, x, t + h, y) dy - \varphi(x) \right]$$
$$= a(t,x)\varphi'(x) + \frac{1}{2}b(t,x)^2\varphi''(x)$$

exists. We obtain

$$\frac{\partial}{\partial t} \int_{-\infty}^{\infty} p(s,x,t,y)\varphi(y)dy$$

$$= \lim_{h \to 0} \frac{1}{h} \left[\int_{-\infty}^{\infty} p(s,x,t+h,y)\varphi(y)dy - \int_{-\infty}^{\infty} p(s,x,t,z)\varphi(z)dz \right]$$

$$= \int_{-\infty}^{\infty} p(s,x,t,z) \lim_{h \to 0} \frac{1}{h} \left[\int_{-\infty}^{\infty} p(t,z,t+h,y)\varphi(y)dy - \varphi(z) \right] dz$$

$$= \int_{-\infty}^{\infty} p(s,x,t,z) \left[a(t,z)\varphi'(z) + \frac{1}{2}b(t,z)^2\varphi''(z) \right] dz.$$

By integrating the last expression in parts, we get

$$\frac{\partial}{\partial t} \int_{-\infty}^{\infty} p(s,x,t,y)\varphi(y)dy = \int_{-\infty}^{\infty} \left[\frac{\partial}{\partial t} p(s,x,t,y) \right] \varphi(y)dy$$

$$= \int_{-\infty}^{\infty} \left\{ -\frac{\partial}{\partial y} [a(t,y)p(s,x,t,y)] \right.$$

$$\left. + \frac{1}{2} \frac{\partial^2}{\partial y^2} [b(t,y)^2 p(s,x,t,y)] \right\} \varphi(y)dy.$$

Hence, equation (11.9) results, because the function $\varphi(y)$ is arbitrary. The theorem is proved. ◻

Equation (11.7) is called the *backward Kolmogorov equation* and (11.9) the *forward Kolmogorov equation*.

Section 12
Linear Stochastic Differential Equations and Linear Random Processes

We know that the general solution of the linear differential equation

(12.1) $\qquad x^{(n)}(t) - a_1(t)x^{(n-1)}(t) - \cdots - a_n(t)x(t) = 0, \quad t > t_0$

(with constant coefficients) can be written in the form

(12.2) $\qquad x(t) = \sum_{k=0}^{n-1} \omega_k(t,t_0)x_k, \quad t \geqslant t_0,$

where we denote by $x_0, ..., x_{n-1}$ the initial values

$$x_0 = x(t_0), ..., x_{n-1} = x^{(n-1)}(t_0)$$

and by $\omega_k(t,t_0)$ the special solutions with initial value $x_k = 1$, $x_j = 0$, for $j \neq k$.

Problem: We consider the Hilbert space H of random variables ξ, $M|\xi|^2 < \infty$. Show the following: For arbitrary random variables $\xi_0, ..., \xi_{n-1} \in H$, the random variable

(12.3) $\qquad \xi(t) = \sum_{k=0}^{n-1} \omega_k(t,t_0)\xi_k, \quad t \geqslant t_0$

has derivatives $\xi^k(t)$ up to order n (which are continuous in quadratic mean) in the space H. It is the unique solution of the differential equation (written in terms of differentials)

(12.4) $\qquad d\xi^{(n-1)}(t) - a_1(t)\xi^{(n-1)}(t)dt - \cdots - a_n(t)\xi(t)dt = 0, \quad t > t_0$

with initial conditions

(12.5) $\qquad \xi(t_0) = \xi_0, ..., \xi^{(n-1)}(t_0) = \xi_{n-1}.$

We assume that for any $\eta \in H$, the scalar function $x(t) = (\xi(t), \eta)$ is the unique solution of the equation (12.1) with initial conditions $x^{(k)}(t_0) = (\xi_k, \eta)$, $k = 0, ..., n - 1$.

We consider the homogeneous analogue of the differential equation (12.4), written in terms of stochastic differentials

(12.6)
$$d\xi^{(n-1)}(t) - a_1(t)\xi^{(n-1)}(t)dt - \cdots - a_n(t)\xi(t)dt$$
$$= b(t)d\eta(t), \quad t > t_0,$$

where we are dealing with the random function $\xi(t)$, $t \geqslant t_0$, in H with continuous derivatives $\xi^{(k)}(t)$ up to order $n - 1$, for which $\xi^{(n-1)}(t)$ has a stochastic differential of the form

(12.7) $\quad d\xi^{(n-1)}(t) = [a_1(t)\xi^{(n-1)}(t) + \cdots + a_n(t)\xi(t)]dt + b(t)d\eta(t)$,

and the derivatives of lower order have, of course, the stochastic differentials

(12.8) $\quad d\xi^{(k)}(t) = \xi^{(k+1)}(t)dt, \quad k = 0, ..., n - 2.$

Recall that the random function $\eta(t)$, $t \geqslant t_0$, on the right side of (12.6) (the standard Wiener process) is not differentiable; we take $d\eta(t)$ as a stochastic measure, for which we defined the stochastic Ito integral.

The difference of any two solutions of (12.6) satisfies the homogeneous equation (12.4) with initial condition zero and, hence, this difference is identically equal to 0. Therefore, the solution $\xi(t)$, $t \geqslant t_0$, of equation (12.6) with initial conditions (12.5) is unique.

If we take in (12.6) the solution $\xi(t)$, $t \geqslant t_0$, with initial conditions

(12.9) $\quad \xi(t_0) = 0, ..., \xi^{(n-1)}(t_0) = 0$

and if we add the solution (12.3) of the homogeneous equation (12.4), then clearly the sum gives us a solution to equation (12.6) with initial condition (12.5).

Theorem. *The solution of the stochastic differential equation (12.6) with initial condition zero is given by the formula*

(12.10) $\quad \xi(t) = \int_{t_0}^{t} \omega(t,s)b(s)d\eta(s), \quad t \geqslant t_0,$

where the function $\omega(t,s)$ of the parameter $t \geqslant s$, s fixed, denotes the solution of the corresponding ordinary differential equation (12.1) with initial conditions

(12.11) $\quad \omega(s,s) = 0, ..., \omega^{(n-2)}(s,s) = 0, \quad \omega^{(n-1)}(s,s) = 1.$

Proof: According to the general formula (9.19), the random function (12.10) has a stochastic differential of the form

$$d\xi(t) = \left[\int_{t_1}^{t} \omega^{(1)}(t,s)b(s)d\eta(s) \right] dt + \omega(t,t)b(t)d\eta(t)$$

for $n > 1$, where $\omega(t,t) = 0$, and, hence, the derivative in quadratic mean exists. It is given by

$$\xi^{(1)}(t) = \int_{t_0}^{t} \omega^{(1)}(t,s)b(s)d\eta(s), \quad t \geqslant t_0.$$

The existence of all $(n - 1)$ derivatives

(12.12) $\qquad \xi^{(k)}(t) = \int_{t_0}^{t} \omega^{(k)}(t,s)b(s)d\eta(s), \quad t \geqslant t_0, \quad k \leqslant n - 1$

can be shown analogously. Using the general formula (9.19) for the $(n-1)$th derivative, we obtain

$$d\xi^{(n-1)}(t) = \left[\int_{t_0}^{t} \omega^{(n)}(t,s)b(s)d\eta(s) \right] dt + \omega^{(n-1)}(t,t)b(t)d\eta(t),$$

where $\omega^{(n-1)}(t,t) = 1$,

$$\omega^{(n)}(t,s) = a_1(t)\omega^{(n-1)}(t,s) + \cdots + a_n(t)\omega(t,s), \quad t > s,$$

which gives together with (12.12) the equations (12.7) - (12.8) for the derivatives $\xi^{(k)}(t)$. Thus, the theorem is proved. $\qquad \square$

We repeat that our definition of the stochastic differential -- cf. (9.15) -- only refers to non-anticipating random functions with respect to some field of events B^t, $t \geqslant t_0$. Obviously, formula (12.10) leads to non-anticipating random function $\xi(t)$, $t \geqslant t_0$, and the same may be said of formula (12.3), if the initial values ξ_k, $k = 0, ..., n - 1$ are non-anticipating (i.e., they are measurable with respect to the σ-algebra of events B^{t_0}.

With these results, we are able to characterize the behaviour of a random process $\xi(t)$, $t \geqslant t_0$, which satisfies the linear stochastic differential equation (12.6) in the following way: For $t \geqslant s$, and without "interval noise", which appears on the right side of (12.6) as $b(t)d\eta(t)$, the trajectory of the process is

$$x(t) = \sum_{k=0}^{n-1} \omega_k(t,s)\xi^{(k)}(s), \quad t \geqslant s,$$

a deterministic function, which is defined by the initial variables $\xi^{(k)}(s)$, $k = 0, ..., n - 1$, where $\omega_k(t,s)$, for $t > s$, is the solution of the ordinary differential equation (12.1) with initial conditions

$$\frac{d^k}{dt^k} \omega_k(s,s) = 1, \qquad \frac{d^j}{dt^j} \omega_k(s,s) = 0$$

for $j \neq k$, $j = 0, ..., n - 1$.

If the term $b(t)d\eta(t)$ is present in the equation, the deviation of the process from the trajectory $x(t)$ is given by the variable

(12.13) $\Delta(t,s) = \xi(t) - x(t) = \int_s^t \omega(t,u)b(u)d\eta(u), \quad t \geq s.$

Let us consider more exactly a (real) random process $\xi(t)$, $t \geq t_0$, which is described by a stochastic differential equation of first order

(12.14) $d\xi(t) = a(t)\xi(t)dt + b(t)d\eta(t)$

with real coefficients $a(t)$, $b(t)$. (It is of the type of stochastic equations that we described in Sec. 10.)

Problem: Using formula (12.13), which holds for all $s \geq t_0$, show that the solution of equation (12.14) is a Markov process with transition density

$$p(s,x,t,y) = \frac{1}{\sqrt{2\pi\sigma^2}} e^{-(y-a)^2/2\sigma^2}, \quad -\infty < y < \infty,$$

where

$$a = \omega(t,s)x, \quad \sigma^2 = \int_s^t [\omega(t,u)b(u)]^2 du.$$

Hint: Use the fact that the stochastic integral,

$$\Delta(t,s) = \int_s^t \omega(t,u)b(u)d\eta(u)$$
$$= \lim_{k\to\infty} \sum_k \omega(t,t_{k-1})b(t_{k-1})[\eta(t_k) - \eta(t_{k-1})],$$

which is defined in terms of the increments of Brownian motion, is a normally distributed random variable.

Problem: (*continuation*). Show that $\xi(t)$, $t \geq t_0$, is a diffusion process with parameters

$$a(s,x) = a(s) \cdot x, \quad b(s,x) = b(s).$$

Hint: Prove that the transition density $p(s,x,t,y)$ satisfies the conditions (11.4) - (11.6).

Problem: Show that the solution of equation (12.14) with initial condition $\xi(t_0) = 0$ is a random process with expectation $M\xi(t) = 0$ and variance $D(t) = M\xi(t)^2$, which is considered to be a function of $t \geq t_0$, namely the solution of the differential equation

(12.15) $\dfrac{d}{dt} D(t) = 2a(t)D(t) + b(t)^2, \quad t > t_0,$

with initial condition $D(t_0) = 0$.

Hint: Take the derivative of the function

$$D(t) = \int_{t_0}^{t} [\omega(t,s)b(s)]^2 ds.$$

In the expression thus obtained set

$$\frac{d}{dt}\omega(t,s) = a(t)\omega(t,s), \quad \omega(t,t) = 1.$$

We shall call an arbitrary random process $\xi(t)$, $t \geqslant t_0$, *linear*, if we bring it into the form

(12.16) $\qquad \xi(t) = \int_{t_0}^{t} \omega(t,s)\eta(ds), \quad t \geqslant t_0,$

where $\omega(t,s)$ is a non-random function

$$\int_{t_0}^{t} |\omega(t,s)|^2 ds < \infty \,,$$

and $\eta(dt)$ is an arbitrary standard stochastic measure with orthogonal values:

$$\mathbf{M}\eta(dt) = 0, \quad \mathbf{M}|\eta(dt)|^2 = dt.$$

We shall call $\omega(t,s)$, $t \geqslant s$, *weight function*.

A linear random process may be obtained, for example, as the solution of a linear stochastic differential equation of nth order with initial condition zero (cf. (12.10)).

We call a linear process *homogeneous*, if its weight function $\omega(t,s)$ only depends on the difference $t - s$:

$$\omega(t,s) = \omega(t - s), \quad t \geqslant s.$$

We shall call the corresponding *weight function* $\omega(t)$, $t \geqslant 0$, *stable*, if

(12.17) $\qquad \int_{0}^{\infty} |\omega(t)|^2 dt < \infty \,.$

Example: (*Linear differential equations with constant coefficients*). We obtain a homogeneous process, if we consider the general linear stochastic differential equation (12.7) with constant coefficients, say

$$a_k(t) = a_k, \quad k = 1, ..., n; \quad b(t) = 1.$$

In fact, suppose that all roots of the characteristic polynome

$$P(z) = z^n - a_1 z^{n-1} - \cdots - a_{n-1} z - a_n$$

lie in the left half-plane $\text{Re}(z) < 0$ of the complex parameter z. The corresponding weight function $\omega(t)$ is the solution of the differential equation

(12.18) $\omega^{(n)}(t) - a_1\omega^{(n-1)}(t) - \cdots - a_n\omega(t) = 0$

with initial conditions

$$\omega^{(n-1)}(0) = 1, \quad \omega^{(k)}(0) = 0, \quad k < n - 1$$

(cf. (12.10)). If the conditions for the solutions of the polynome $P(z)$ hold, the function $\omega(t)$ exponentially decreases with $t \to \infty$, so that it satisfies condition (12.17). Setting $\omega(t) = 0$ for $t < 0$, we introduce the well-known *Fourier transformation* formula:

(12.19) $\displaystyle\int_0^\infty e^{i\lambda t}\omega(t)dt = \frac{1}{P(-i\lambda)}, \quad -\infty < \lambda < \infty.$

This formula can easily be obtained by partial integration of

$$\int_0^\infty e^{i\lambda t}[\omega^{(n)}(t) - a_1\omega^{(n-1)}(t) - \cdots - a_n\omega(t)]dt = 0.$$

Let us consider the general homogeneous linear process

(12.20) $\displaystyle\xi(t) = \int_{t_0}^t \omega(t-s)\eta(ds), \quad t \geqslant t_0$

with a stable weight function and its behaviour after a large time interval $t - t_0 \to \infty$. Formally, it is easier to assume that $t_0 \to -\infty$ (we suppose that the stochastic measure $\eta(dt)$ is given on the real axis $-\infty < t < \infty$). We set $\omega(t) = 0$ for $t < 0$ and consider the random processs ξ

(12.21) $\displaystyle\xi^*(t) = \int_{-\infty}^\infty \omega(t-s)\eta(ds) = \int_{-\infty}^t \omega(t-s)\eta(ds), \quad -\infty < t < \infty.$

It has the expectation $M\xi^*(t) = 0$ and the *correlation function*

$$B(t,s) = M\xi^*(t)\overline{\xi^*(s)}$$

$$= \int_{-\infty}^\infty \omega(t-u)\overline{\omega(s-u)}du = \int_{-\infty}^\infty \omega(t-s+u)\overline{\omega(u)}\,du$$

$$= B(t-s), \quad -\infty < s, t < \infty,$$

which depends only on the difference $t - s$ (a random process of this type is called *stationary in the wide sense*; cf. Sec. 13). Comparing (12.20) and (12.21), we easily obtain

$$\|\xi(t) - \xi^*(t)\|^2 = \left\|\int_{-\infty}^{t_0} \omega(t-s)\eta(ds)\right\|^2$$

$$= \int_{-\infty}^{t_0} |\omega(t-s)|^2 ds = \int_{t-t_0}^\infty |\omega(u)|^2 du \to 0$$

for $t - t_0 \to \infty$. Hence, we derive the following

Theorem. *For $t - t_0 \to \infty$, the homogeneous linear process (12.20) with stable weight function converges in quadratic mean to the process (12.21), which is stationary in the wide sense.*

Section 13
Stationary Processes
Spectral Analysis and Linear Transformations

The random process $\xi(t)$, $\mathbf{M}|\xi(t)|^2 < \infty$, on the real line $-\infty < t < \infty$ is called *stationary in the wide sense,* if its expectation $\mathbf{M}\xi(t)$ does not depend on t (we set $\mathbf{M}\xi(t) = 0$), and if the *correlation function*

$$B(t,s) = \mathbf{M}\xi(t)\overline{\xi(s)}\,, \quad -\infty < s, \ t < \infty,$$

depends only on the difference $t - s$:

$$(13.1) \qquad \mathbf{M}\xi(t)\overline{\xi(s)} = B(t - s).$$

The corresponding function $B(t)$, $-\infty < t < \infty$, is also called *correlation function* of the stationary process.

The property of stationarity in the wide sense expresses the fact that the correlation of the variables $\xi(t_1)$, $\xi(t_2)$ for arbitrary t_1, t_2 does not change under time shift, i.e., the variables $\xi(t_1 + t)$, $\xi(t_2 + t)$ have the same correlation for all $t \geqslant 0$.

As one of the simplest examples of a process of this type may serve the harmonic oscillation $\xi(t) = \alpha \exp(i(\lambda t + \theta))$, $-\infty < t < \infty$, of frequency λ with random amplitude $|\alpha|$ and phase θ, where α and θ are independent real random variables with $\mathbf{M}\alpha = 0$, $\mathbf{M}|\alpha|^2 < \infty$.

Problem: Let the random variable θ of the above mentioned example be uniformly distributed on the interval $-\pi \leqslant \theta \leqslant \pi$. Then show that the probability distribution of the variables $\xi(t_1)$, ..., $\xi(t_n)$ for arbitrary t_1, ..., t_n is invariant under time shift, i.e., the variables $\xi(t_1 + t)$, ..., $\xi(t_n + t)$[5] are characterized by the same probability distribution for any $t \geqslant 0$. A random process of this property is called *strictly stationary.*

---

[5] In the case of probability distributions of complex variables of the form $\xi = \xi_1 + i\xi_2$, the probability distributions of their real components (ξ_1, ξ_2) are considered separately.

We shall consider processes that are stationary in the wide sense, which have a spectral representation of the form

(13.2) $\xi(t) = \int_{-\infty}^{\infty} e^{i\lambda t}\varphi(\lambda)\zeta(d\lambda), \quad -\infty < t < \infty,$

where $\zeta(d\lambda)$ is a standard stochastic measure with orthogonal values on the real line $-\infty < \lambda < \infty$:

(13.3) $\mathbf{M}\zeta(d\lambda) = 0, \quad \mathbf{M}|\zeta(d\lambda)|^2 = d\lambda;$

The non-random function $\varphi(\lambda)$ satisfies the condition $\int_{-\infty}^{\infty}|\varphi(\lambda)|^2 d\lambda < \infty$ under which the stochastic integral (13.2) is defined for each t. The expression (13.2) represents the harmonic oscillation

$\varphi(\lambda)e^{i\lambda t}, \quad -\infty < t < \infty$

of frequency λ with "weight" $\varphi(\lambda)$;

(13.4) $f(\lambda) = |\varphi(\lambda)|^2$

as a function of the parameter λ, $-\infty < \lambda < \infty$, is called *spectral density*. It characterizes the weight of the different harmonic components of the random process (13.2), depending on the frequency λ. According to the general formulas (8.16) and (8.12) for stochastic integrals, we have

$\mathbf{M}\xi(t) = 0,$

$\mathbf{M}\xi(t)\overline{\xi(s)} = \int_{-\infty}^{\infty} e^{i\lambda(t-s)}|\varphi(\lambda)|^2 d\lambda,$

so that each random process (13.2) is stationary in the wide sense, and its correlation function is

(13.5) $B(t) = \int_{-\infty}^{\infty} e^{i\lambda t} f(\lambda) d\lambda, \quad -\infty < t < \infty.$

Setting formally $\Phi(d\lambda) = \varphi(\lambda)\zeta(d\lambda)$, we shall use the spectral representation (13.2) in the form

(13.6) $\xi(t) = \int_{-\infty}^{\infty} e^{i\lambda t}\Phi(d\lambda), \quad -\infty < t < \infty.$

This representation is useful, if we consider *linear transformations* of a random process $\xi(t)$ of the type

(13.7) $\eta(t) = \int_{-\infty}^{\infty} e^{i\lambda t}\psi(\lambda)\Phi(d\lambda), \quad -\infty < t < \infty,$

which obviously transforms the harmonic components of the initial process by giving them the corresponding "weight" $\psi(\lambda)$, depending on the frequency λ. This transformation can give more weight to some components and less weight to others for some λ. Of course, the *weight function* $\psi(\lambda)$ has to fulfill the condition

(13.8) $\int_{-\infty}^{\infty}|\psi(\lambda)|^2 f(\lambda)d\lambda < \infty$,

under which the stochastic integral on the right side of (13.7) is defined for each t.

Let us consider some examples of linear transformations of a stationary process $\xi(t)$ of type (13.6).

Example: (*Differentiation*) Let the spectral density $f(\lambda)$ be such that

$$\int_{-\infty}^{\infty}|\lambda|^2 f(\lambda)d\lambda < \infty .$$

Then the stationary process $\xi(t)$ has a derivative in quadratic mean:

$$\xi'(t) = \lim_{h\to 0}\frac{\xi(t+h) - \xi(t)}{h}$$

$$= \int_{-\infty}^{\infty}\left[\lim_{h\to 0}\frac{e^{i\lambda(t+h)} - e^{i\lambda t}}{h}\right]\Phi(d\lambda) = \int_{-\infty}^{\infty}e^{i\lambda t}(i\lambda)\Phi(d\lambda).$$

Example: (*Integration*). Let the function $c(t)$ be integrable and set

$$\psi(\lambda) = \int_{-\infty}^{\infty}e^{-i\lambda t}c(t)dt.$$

Then

$$\eta(t) = \int_{-\infty}^{\infty}e^{i\lambda t}\psi(\lambda)\Phi(d\lambda) = \int_{-\infty}^{\infty}e^{i\lambda t}\left[\int_{-\infty}^{\infty}e^{-i\lambda s}c(s)ds\right]\Phi(d\lambda)$$

$$= \int_{-\infty}^{\infty}c(s)\left[\int_{-\infty}^{\infty}e^{i\lambda(t-s)}\Phi(d\lambda)\right]ds$$

$$= \int_{-\infty}^{\infty}c(s)\xi(t - s)ds = \int_{-\infty}^{\infty}c(t - s)\xi(s)ds.$$

Example: (*Low frequency filter, estimation of the expectation*). Set

$$\psi(\lambda) = \frac{1}{T}\int_{0}^{T}e^{-i\lambda t}dt = \frac{e^{-i\lambda T} - 1}{-i\lambda T} .$$

The corresponding linear transformation is

$$\eta(t) = \int_{-\infty}^{\infty}e^{i\lambda t}\psi(\lambda)\Phi(d\lambda) = \frac{1}{T}\int_{t-T}^{t}\xi(s)ds, \quad -\infty < t < \infty .$$

Here the low frequent components remain nearly unchanged, whereas the components with frequency $|\lambda| > \epsilon$, where ϵ becomes small with T becoming large, are practically suppressed. This transformation may be used for the estimation of the unknown constant θ by the "observed"

$$x(t) = \theta + \xi(t), \quad 0 \leqslant t \leqslant T.$$

For the *empirical expectation*

$$\hat{\theta} = \frac{1}{T} \int_0^T x(t)dt = \theta + \frac{1}{T} \int_0^T \xi(t)dt$$

we have in fact, with bounded spectral density $f(\lambda)$, that

$$\|\hat{\theta} - \theta\|^2 = \left\| \frac{1}{T}\int_0^T \xi(t)dt \right\|^2$$

$$= \int_{-\infty}^{\infty} |\psi(\lambda)|^2 f(\lambda)d\lambda = \frac{1}{T}\int_{-\infty}^{\infty} \left| \frac{e^{i\mu}-1}{i\mu} \right|^2 f\left(\frac{\mu}{T}\right) d\mu$$

$$\leqslant \frac{C}{T} \to 0$$

for $T \to \infty$ (cf. the law of large numbers).

Let us consider the stationary linear process of the form

(13.9) $\xi(t) = \int_{-\infty}^{\infty} \omega(t-s)\eta(ds), \quad -\infty < t < \infty,$

where $\eta(dt)$ is the standard stochastic measure with orthogonal values on the time axis $-\infty < t < \infty$:

(13.10) $M\eta(dt) = 0, \quad M|\eta(dt)|^2 = dt,$

and the weight function $\omega(t)$ fulfills the condition

(13.11) $\int_{-\infty}^{\infty} |\omega(t)|^2 dt < \infty .$

We have already treated this type of processes that are stationary in the wide sense (cf. (12.20), (12.21)). On the following pages we show that the stationary process (13.9) has the spectral transformation (13.2), which we obtain by using the *Fourier transformation of the stochastic measure* $\eta(dt)$. With regard to (13.11), it will be convenient to use the L^2 space of functions in $\mathbb{R}$, that are integrable in quadratic mean, with scalar product

$$(u_1, u_2) = \int_{-\infty}^{\infty} u_1(t)\overline{u_2(t)}\, dt, \quad u_1, u_2 \in L^2$$

and norm $\|u\| = (u,u)^{1/2}$, defining the distance in quadratic mean as

$$\|u_1 - u_2\| = \left[\int_{-\infty}^{\infty} |u_1(t) - u_2(t)|^2 dt \right]^{1/2} .$$

Recall[6] that the Fourier Transformation

---

[6]Cf. for example, L. D. Kudryavtsev, Mathematical analysis, II. - Moscow, Vyshaya Shkola, 1970 (Russian).

(13.12) $\tilde{u}(\lambda) = \dfrac{1}{\sqrt{2\pi}} \displaystyle\int_{-\infty}^{\infty} e^{i\lambda t} u(t) dt, \quad -\infty < \lambda < \infty$

of functions $u(t)$, $-\infty < t < \infty$, which are integrable in quadratic mean, is defined as

$$\tilde{u}(\lambda) = \lim_{n \to \infty} \frac{1}{\sqrt{2\pi}} \int_{-n}^{n} e^{i\lambda t} u(t) dt,$$

where convergence in quadratic mean is understood in terms of the functions

$$\tilde{u}_n(\lambda) = \frac{1}{\sqrt{2\pi}} \int_{-n}^{n} e^{i\lambda t} u(t) dt \quad \text{with} \quad n \to \infty :$$

$$\|\tilde{u} - \tilde{u}_n\|^2 = \int_{-\infty}^{\infty} |\tilde{u}(\lambda) - \tilde{u}_n(\lambda)|^2 d\lambda \to 0.$$

This transformation defines an isomorphism of the L^2 space of functions $u(t)$, $-\infty < t < \infty$, which are integrable in quadratic mean on the L^2 space of the functions $\varphi(\lambda)$, $-\infty < \lambda < \infty$, which are integrable in quadratic mean. This application is continuous in the sense of convergence in quadratic mean, and the *inverse Fourier transformation* for $\varphi(\lambda) = \tilde{u}(\lambda)$ is given by the analogue formula to (13.12)

(13.13) $u(t) = \dfrac{1}{\sqrt{2\pi}} \displaystyle\int_{-\infty}^{\infty} e^{-i\lambda t} \varphi(\lambda) d\lambda, \quad -\infty < t < \infty .$

Most important for us is the well-known equation

(13.14) $(\tilde{u}_1, \tilde{u}_2) = \displaystyle\int_{-\infty}^{\infty} \tilde{u}_1(\lambda) \cdot \overline{\tilde{u}_2(\lambda)} d\lambda = \int_{-\infty}^{\infty} u_1(t) \cdot \overline{u_2(t)} \, dt = (u_1, u_2),$

which holds for any function $u_1, u_2 \in L^2$. For $u_1 = u_2$ it is the so-called *Plancherel* equation:

(13.15) $\|\tilde{u}\|^2 = \displaystyle\int_{-\infty}^{\infty} |\tilde{u}(\lambda)|^2 d\lambda = \int_{-\infty}^{\infty} |u(t)|^2 dt = \|u\|^2.$

Recall that we can obtain equation (13.14) it we apply the formula of the inverse application (13.13) to the *convolution*

$$u(t) = \frac{1}{\sqrt{2\pi}} \int_{-\infty}^{\infty} u_1(t + s) \cdot \overline{u_2(s)} \, ds.$$

If we suppose that the functions $u_1(t)$, $u_2(t)$ are integrable, then

$$\tilde{u}_1(\lambda) \cdot \overline{\tilde{u}_2(\lambda)}$$

$$= \frac{1}{\sqrt{2\pi}} \int_{-\infty}^{\infty} e^{-i\lambda s} \left[\frac{1}{\sqrt{2\pi}} \int_{-\infty}^{\infty} e^{i\lambda(t+s)} u_1(t + s) ds \right] \overline{u_2(s)} \, ds$$

$$= \frac{1}{\sqrt{2\pi}} \int_{-\infty}^{\infty} e^{i\lambda t} u(t) dt;$$

it is obvious that $\bar{u}_1(\lambda) \cdot \overline{\bar{u}_2(\lambda)} = \bar{u}(\lambda)$, and by the inversion formula (13.13) we have

$$\int_{-\infty}^{\infty} e^{-i\lambda t}\,\bar{u}_1(\lambda) \cdot \overline{\bar{u}_2(\lambda)}d\lambda = \int_{-\infty}^{\infty} u_1(t+s) \cdot \overline{u_2(s)}\,ds,$$

which gives in case $t = 0$ equation (13.14) and in case $u_1 = u_2$ the Plancherel equation. We can take arbitrary $u_1, u_2 \in L^2$ instead of integrable functions in L^2, if we apply a limiting procedure using convergence in quadratic mean.

We need the inverse Fourier transformation of the indicator function

$$\varphi(\lambda) = 1_\Delta(\lambda) = \begin{cases} 1, & \lambda \in \Delta, \\ 0, & \lambda \notin \Delta \end{cases}$$

on the half-interval $\Delta = (\lambda_1, \lambda_2]$, which is

$$(13.16) \qquad u_\Delta(t) = \frac{1}{\sqrt{2\pi}}\int_{-\infty}^{\infty} e^{-i\lambda t}1_\Delta(\lambda)d\lambda = \frac{1}{\sqrt{2\pi}}\frac{e^{-i\lambda_2 t} - e^{-i\lambda_1 t}}{-it}.$$

Returning to the stochastic measure $\eta(dt)$ in (13.10), we set

$$(13.17) \qquad \zeta(\Delta) = \frac{1}{\sqrt{2\pi}}\int_{-\infty}^{\infty} \frac{e^{-i\lambda_2 t} - e^{-i\lambda_1 t}}{-it}\,\eta(dt), \qquad \Delta = (\lambda_1, \lambda_2],$$

which is an additive function on all possible half-intervals $\Delta = (\lambda_1, \lambda_2]$. Obviously, with the general equations (13.14), (13.15) holding, this is an additive function with orthogonal values, which defines a stochastic measure of the type (13.3) on the real line $-\infty < \lambda < \infty$. According to formulas (8.12), (8.16), for stochastic integrals we have indeed

$$\mathbf{M}\zeta(\Delta) = 0,$$

$$\mathbf{M}\zeta(\Delta_1)\overline{\zeta(\Delta_2)} = \int_{-\infty}^{\infty} u_{\Delta_1}\overline{u_{\Delta_2}(t)}\,dt$$

$$= \int_{-\infty}^{\infty} 1_{\Delta_1}(\lambda) \cdot \overline{1_{\Delta_2}(\lambda)}d\lambda = \int_{\Delta_1 \cap \Delta_2} d\lambda.$$

We shall prove now that the equation

$$(13.18) \qquad \int_{-\infty}^{\infty} \varphi(\lambda)\zeta(d\lambda) = \int_{-\infty}^{\infty} u(t)\eta(dt)$$

holds for any function $\varphi(\lambda)$ and $u(t)$, related by the Fourier-transformation $\varphi(\lambda) = \bar{u}(\lambda)$. In fact, for indicator functions $\varphi(\lambda) = 1_\Delta(\lambda)$, the equation appears to be the initial formula (13.17), which defines the stochastic measure $\zeta(d\lambda)$. Obviously, equation (13.18) can be extended from indicator functions to the linear combinations

(13.19) $\varphi(\lambda) = \sum_k c_k 1_{\Delta_k}(\lambda).$

It is possible to approximate in quadratic mean an arbitrary function $\varphi \in L^2$ by functions $\varphi_n(\lambda)$ of the type (13.19):

$$\|\varphi - \varphi_n\|^2 = \int_{-\infty}^{\infty} |\varphi(\lambda) - \varphi_n(\lambda)\|^2 d\lambda \to 0,$$

which allows to generalize equation (13.18) from functions $\varphi = \varphi_n$ of type (13.19) to arbitrary functions $\varphi \in L^2$.

Applying a well known property of stochastic integrals (cf. (8.13)), we indeed obtain

$$\int_{-\infty}^{\infty} \varphi(\lambda)\zeta(d\lambda) = \lim_{n \to \infty} \int_{-\infty}^{\infty} \varphi_n(\lambda)\zeta(d\lambda)$$

$$= \lim_{n \to \infty} \int_{-\infty}^{\infty} u_n(t)\eta(dt) = \int_{-\infty}^{\infty} u(t)\eta(dt),$$

where for $\varphi_n(\lambda) = \tilde{u}_n(\lambda)$

$$\|u - u_n\|^2 = \int_{-\infty}^{\infty} |u(t) - u_n(t)|^2 dt = \int_{-\infty}^{\infty} |\varphi(\lambda) - \varphi_n(\lambda)|^2 dt \to 0.$$

Now we turn to the stationary linear process (13.19) and apply the general formula (13.18). We get

$$\xi(t) = \int_{-\infty}^{\infty} \omega(t - s)\eta(ds) = \int_{-\infty}^{\infty} e^{i\lambda t}\varphi(\lambda)\zeta(d\lambda),$$

where

(13.20) $\varphi(\lambda) = \dfrac{1}{\sqrt{2\pi}} \int_{-\infty}^{\infty} e^{-i\lambda t}\omega(t)dt,$

and the function $\exp(i\lambda t) \cdot \varphi(\lambda)$ is the Fourier transformation of the function $u(s) = \omega(t - s)$, $-\infty < s < \infty$.

The result can be generalized in the following theorem:

Theorem. *A stationary linear process of the form* (13.10) *has a spectral representation* (13.2) *with weight function* (13.20).

With formula (13.20), we can easily obtain the spectral representation of stationary processes, which are described by linear stochastic differential equations with constant coefficients. In fact, if we deal with a linear stationary process (13.10) with weight function $\omega(t)$, which is the solution of the differential equation (13.18) with characteristic polynome $P(z)$, which has all roots in the left half-plane Re $z < 0$, then the weight function $\varphi(\lambda)$ in the spectral representation (13.2) of our linear stationary process is

(13.21) $\varphi(\lambda) = \dfrac{1}{\sqrt{2\pi}\, P(i\lambda)} .$

Problem: Consider the differential equation

$$\omega''(t) + 2h\omega'(t) + a^2\omega(t) = 0$$

and the process (13.9) together with conditions (13.10) and (13.11). For an ω that satisfies the equation above, this process describes a stabilizing (random) oscillating pendulum, the displacement of which is given by the equation above. Find the spectrum of this oscillation, or, put more precisely, find the corresponding spectral density $f(\lambda)$.

Section 14
Some Problems of Optimal Estimation

Let us consider the model of a random process $\xi(t)$, $t \geqslant 0$, with stochastic differential

(14.1) $d\xi(t) = \theta(t)dt + d\eta(t).$

We suppose that $\theta(t) = \theta$ is an unknown constant, which has to be estimated with the help of the values $\xi(t)$, $0 \leqslant t \leqslant T$. Assume, we are given the random process $\xi(t)$, $t \geqslant 0$, but we do not know the drift $\theta t = M\xi(t)$, $t \geqslant 0$. The deviation of ξ from the expectation $M\xi(t)$ is described by the standard Wiener process $\eta(t) = \xi(t) - \theta t$, $t \geqslant 0$.

We must estimate the unknown parameter θ from the "observed" trajectory of the process $\xi(t)$, $t \geqslant 0$ (Fig. 7). We may take as an estimator $\hat{\theta}$ of the unknown parameter θ the "weighted average"

(14.2) $\hat{\theta} = \int_0^T c(t)d\xi(t) = \int_0^T c(t)\theta(t)dt + \int_0^T c(t)d\eta(t)$

with weight function $c(t)$,

(14.3) $\int_0^T c(t)dt = 1.$

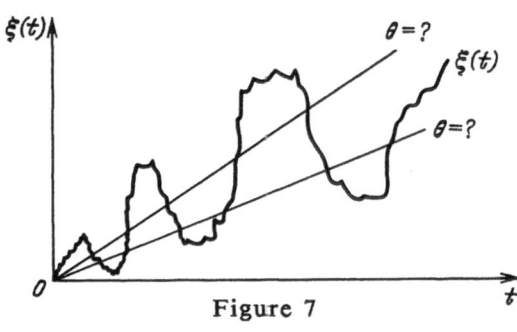

Figure 7

Of course, this function must be integrable in quadratic mean:

$$\int_0^T |c(t)|^2 dt < \infty \, .$$

Only for such functions is the stochastic integral of (14.2) defined for the standard stochastic measure $d\eta(t)$:

$$\mathbf{M}\eta(dt) = 0, \quad \mathbf{M}|\eta(dt)|^2 = dt.$$

With condition (14.3), our estimator is

$$\hat{\theta} = \theta + \int_0^T c(t) d\eta(t)$$

and

(14.4) $\mathbf{M}\hat{\theta} = \theta$

for all θ. An estimator, that fulfills this requirement, is called unbiased. The error in quadratic mean $\|\hat{\theta} - \theta\| = (\mathbf{M}|\hat{\theta} - \theta|^2)^{1/2}$ of the estimation of the unknown parameter θ by the estimator $\hat{\theta}$ can easily be calculated as

$$\|\hat{\theta} - \theta\|^2 = \mathbf{M}\left|\int_0^T c(t) d\eta(t)\right|^2 = \int_0^T |c(t)|^2 dt.$$

In a natural way, we define the *optimal estimator* of all these estimators $\hat{\theta}$ as the one, for which

(14.5) $\|\hat{\theta} - \theta\| = \min.$

Such an estimator exists, and it can easily be found. In fact, the minimum (14.5) is attained for the weight function

(14.6) $c^0(t) = 1/T, \quad 0 \leqslant t \leqslant T,$

since we have, with $\Delta(t) = c(t) - c^0(t)$

$$\int_0^T c^0(t)\Delta(t) dt = 0$$

and

$$\int_0^T |c(t)|^2 dt = \int_0^T |c^0(t) + \Delta(t)|^2 dt$$

$$= \int_0^T |c^0(t)|^2 dt + \int_0^T |\Delta(t)|^2 dt \geqslant \int_0^T |c^0(t)|^2 dt.$$

Problem: Find the optimal unbiased estimator (14.2) for the unknown constant θ (14.1) with the function $\theta(t) = \theta \cdot f(t)$, where $f(t)$ is an arbitrary (real) function,

$$\int_0^T |f(t)|^2 dt < \infty .$$

Hint: Instead of (14.3) apply the condition

(14.7) $\int_0^T c(t)f(t)dt = 1$

and show that the weight function of the optimal estimator is

(14.8) $c^0(t) = \dfrac{1}{\displaystyle\int_0^T |f(t)|^2 dt} f(t), \quad 0 \leqslant t \leqslant T.$

Problem: In (14.1) let

(14.9) $\theta(t) = \sum_{k=1}^{n} \theta_k f_k(t),$

where the functions $f_1(t)$, ..., $f_n(t)$ are taken such that

$$\int_0^T f_k(t)f_j(t)dt = \begin{cases} 1, & k = j, \\ 0, & k \neq j. \end{cases}$$

Find the optimal unbiased estimator $\hat{\theta} = \hat{\theta}_1$, ..., $\hat{\theta}_n$ of type (14.2) for the arbitrary constants $\theta = \theta_1$, ..., θ_n.

Hint: Instead of (14.7) apply the condition that the corresponding estimators

$$\hat{\theta}_k = \int_0^T c_k(t)d\xi(t) = \int_0^T c_k(t)\theta(t)dt + \int_0^T c_k(t)d\eta(t)$$

are unbiased:

(14.10) . $\int_0^T c_k(t)f_j(t)dt = \begin{cases} 1, & k = j, \\ 0, & k \neq j, \end{cases}$

and show that the weight functions of the optimal estimators are

(14.11) $c_k^0(t) = f_k(t), \quad k = 1, ..., n.$

Let us now look at some general approaches to other problems of estimation.

Suppose we have to estimate the (real) random variable ξ, $M|\xi|^2 < \infty$, with the help of the variables $\eta_1, ..., \eta_n$. Take as an estimator

(14.12) $\eta = \varphi(\eta_1, ..., \eta_n),$

where $\varphi(y_1, ..., y_n)$ is some (real) function of the parameters $y_1, ..., y_n$; then the estimator (14.12) is defined as a function of the variables

$\eta_1, \ldots, \eta_n$. Let us consider all possible estimators η, $M|\eta|^2 < \infty$, saying that the estimator η is as better as the *error in quadratic mean*

$$\|\xi - \eta\| = (M \|\xi - \eta\|^2)^{1/2}$$

is smaller.

Naturally, the question arises how to find the *optimal estimator* η^0, which has the smallest error in quadratic mean for ξ estimated:

$$\|\xi - \eta^0\| = \min \|\xi - \eta\|,$$

where the minimum is taken over all estimators of the form (14.12). To solve this problem, we return to the conditional probability distribution

$$P_\xi(dx|y_1, \ldots, y_n), \quad -\infty < y_1, \ldots, y_n < \infty$$

of the variable ξ with respect to $\eta_1, \ldots, \eta_n$ and to the conditional mathematical expectation

$$\varphi^0(y_1, \ldots, y_n) = M(\xi|y_1, \ldots, y_n) = \int_{-\infty}^{\infty} x P_\xi(dx|y_1, \ldots, y_n).$$

Theorem. *The optimal estimator is*

(14.13) $\eta^0 = \varphi^0(\eta_1, \ldots, \eta_n) = M(\xi|\eta_1, \ldots, \eta_n).$

Proof: Let us denote by $H(\eta_1, \ldots, \eta_n)$ the set of all estimators η of the form (14.12), satisfying the condition $M|\eta|^2 < \infty$. We can show that the estimator η^0, defined by formula (14.13), belongs to $H(\eta_1, \ldots, \eta_n)$, i.e., $M|\eta^0|^2 < \infty$. In fact, we have, according to the well-known inequality for the mathematical expectation, that

$$|M(\xi|y_1, \ldots, y_n)|^2 \leqslant M(|\xi|^2|y_1, \ldots, y_n)$$

for fixed $y_1, \ldots, y_n$, and, hence,

$$|M(\xi|\eta_1, \ldots, \eta_n)|^2 \leqslant M(|\xi|^2|\eta_1, \ldots, \eta_n).$$

With the formula of the complete mathematical expectation we find that

$$MM(|\xi|^2|\eta_1, \ldots, \eta_n) = M|\xi|^2 < \infty. \quad \square$$

Let us return to the real Hilbert space H of random variables with the scalar product

$$(\xi_1, \xi_2) = M\xi_1\xi_2, \quad \xi_1, \xi_2 \in H$$

(cf. (7.1)). For each fixed y_1, ..., y_n we have

$$\mathbf{M}(\xi|y_1, ..., y_n) \cdot \varphi(y_1, ..., y_n)$$

$$= \mathbf{M}[\xi \cdot \varphi(y_1, ..., y_n)|y_1, ..., y_n],$$

and, hence, for $\eta = \varphi(\eta_1, ..., \eta_n)$

$$(\eta^0, \eta) = \mathbf{M}[\mathbf{M}(\xi|\eta_1, ..., \eta_n) \cdot \eta]$$

$$= \mathbf{M}[\mathbf{M}(\xi\eta|\eta_1, ..., \eta_n)] = \mathbf{M}\xi\eta = (\xi, \eta),$$

which implies

$$(\xi - \eta^0, \eta) = 0, \quad \eta \in H(\eta_1, ..., \eta_n).$$

We take an arbitrary estimator $\eta \in H(\eta_1, ..., \eta_n)$ and set $\Delta = \eta - \eta^0$. According to the equation above, the difference $\xi - \eta^0$ is the orthogonal variable $\Delta \in H(\eta_1, ..., \eta_n)$, and, hence,

$$\|\xi - \eta\|^2 = \|(\xi - \eta^0) + \Delta\|^2 = \|\xi - \eta^0\|^2 + \|\Delta\|^2,$$

where $\|\Delta\|^2 \geqslant 0$. Obviously, the estimator $\eta = \eta^0$, for which $\Delta = 0$ is optimal.

Geometrically, the variable η^0 is the projection of the variable $\xi \in$ **H** on the linear subspace $H(\eta_1, ..., \eta_n)$ in our Hilbert space **H**.

If we take the Hilbert space **H** and if we apply the fact that for each variable $\xi \in$ **H** and for each (closed) linear subspace $H \subseteq$ **H**, the projection $\hat{\xi} \in H$ of the variable ξ on H exists:

$$(14.14) \qquad \|\xi - \hat{\xi}\| = \min_{\eta \in H} \|\xi - \eta\|.$$

Then we may say in general that $\hat{\xi}$ defines the optimal estimator for ξ among the possible estimators $\eta \in H$. We know that the projection $\hat{\xi}$ is defined by the orthogonality condition

$$(14.15) \qquad (\xi - \hat{\xi}, \eta) = 0, \quad \eta \in H.$$

To verify this for all $\eta \in H$, we take an arbitrary complete system of variables $\eta \in H$.

Problem: (*Linear estimation*). Let η_1, ..., $\eta_n \in$ **H**, and let H be the linear subspace in **H** of all linear estimators $\eta = \sum_{k=1}^{n} c_k \eta_k$ (with arbitrary constants c_1, ..., c_n). Find the optimal linear estimator $\hat{\xi} \in H$.

Hint: Turn to the orthonormal variables η_1, ..., η_n by a linear transformation:

$$(\eta_k, \eta_j) = \begin{cases} 1, & k = j, \\ 0, & k \neq j \end{cases}$$

and show that

$$\hat{\xi} = \sum_{k=1}^{n} (\xi, \eta_k)\eta_k .$$

We shall now deal with the question of the expected evolution of our random process $\xi(t)$, $t \geqslant t_0$, considering it as a function in our Hilbert space **H**. We suppose that $\mathbf{M}|\xi(t)|^2 < \infty$ for all $t \geqslant t_0$ and denote by $H^s = H(\xi(u), u \leqslant s)$ the set of all random variables of **H** which are functions $\eta = \varphi(\xi(s_1), ..., \xi(s_m))$ of the values $\xi(u)$ at some times $u = s_1, ..., s_m$ or their limits in quadratic mean. By definition, H^s is a closed linear subspace in **H**, and the projection of the variable $\xi(t) \in \mathbf{H}$ on this subspace exists. We denote it by

$$\hat{\xi}(s,t) = \mathbf{M}(\xi(t)|\xi(u), \ u \leqslant s)$$

and call it the *optimal prediction* for $\xi(t)$, observing the variables $\xi(u)$, $u \leqslant s$:

$$\|\xi(t) - \hat{\xi}(t)\| = \min_{\eta \in H^s} \|\xi(t) - \eta\|.$$

Problem: Show that the optimal prediction for the Brownian motion is

$$(14.16) \qquad \mathbf{M}(\xi(t)|\xi(u), u \leqslant s) = \xi(s), \ t \geqslant s.$$

Hint: Make use of the fact that the difference $\xi(t) - \xi(s)$ does not depend on the variables $\xi(u), u \leqslant s$. We call a random process $\xi(t), t \geqslant t_0$, a *martingale*, if it satisfies the property (14.16).

Problem: Let the random process $\xi(t)$, $t \geqslant t_0$, be described by the linear stochastic differential equation (12.7). Show that the optimal prediction, given the variables $\xi(u)$, $u \leqslant s$, is

$$(14.17) \qquad \mathbf{M}(\xi(t)|\xi(u), \ u \leqslant s) = \sum_{k=0}^{n-1} \omega_k(t,s)\xi^{(k)}(s), \ t \geqslant s,$$

where $\omega_k(t,s)$, $t \geqslant s$, is the solution of the corresponding ordinary differential equation (12.1) with initial conditions

$$\frac{d^k}{dt^k} \omega_k(s,s) = 1, \quad \frac{d^j}{dt^j} \omega_k(s,s) = 0, \quad j \neq k, \quad j = 0, ..., n-1.$$

Hint: Apply the fact that the difference

$$\Delta(s,t) = \xi(t) - \hat{\xi}(s,t) = \int_s^t \omega(t,u)b(u)d\eta(u), \quad t \geqslant s$$

does not depend on the variable $\xi(u)$, $u \leqslant s$ (cf. (12.13)). Notice that the optimal prediction (14.17) is *linear* in the sense that the variable $\hat{\xi}(s,t) = M(\xi(t) \mid \xi(u), u \leqslant s)$, may be found as the limit in quadratic mean of the linear combinations of the variables $\xi(u)$, $u \leqslant s$, since

$$\xi^{(1)}(s) = \lim_{h \to 0} \frac{\xi(s) - \xi(s - h)}{h}$$

$$\vdots$$

$$\xi^{(n-1)}(s) = \lim_{h \to 0} \frac{\xi^{(n-2)}(s) - \xi^{(n-2)}(s - h)}{h}$$

Above, we defined the optimal estimator by minimizing the error in quadratic mean. This procedure is not applicable in all cases.

Assume we consider a system the state of which is described by the random variable ξ, which takes only one of the possible values $x = x_1, ..., x_n$ with corresponding probability $P_\xi(x)$. With the estimator ξ, given the variable $\eta_1, ..., \eta_n$, the conditional mathematical expectation $M(\xi|\eta_1, ..., \eta_n)$ does not directly give us a hint, in which state $\xi = x_1, ..., x_m$ the system is. What we can say about the state of the system for $M(\xi|\eta_1, ..., \eta_n) \neq (x_1, ..., x_m)$ is uncertain.

Let us consider all possible estimators $\eta = \varphi(\eta_1, ..., \eta_n)$ with values $\eta = x_1, ..., x_m$. We call the estimator η^0 *optimal*, if the probability of error $P\{\eta^0 \neq \xi\}$ is minimal:

(14.18) $P\{\eta^0 \neq \xi\} = \min_\eta P\{\eta \neq \xi\}.$

We use the conditional probability distribution[7]

$$P_\xi(x|\eta_1, ..., \eta_n), \quad x = x_1, ..., x_m$$

of the variable ξ with respect to $\eta_1, ..., \eta_n$ and we set η^0 equal to the value x, for which the conditional probability $P_\xi(x|\eta_1, ..., \eta_n)$ is maximal:

(14.19) $P_\xi(\eta^0|\eta_1, ..., \eta_n) = \max_{x=x_1,...,x_m} P_\xi(x|\eta_1, ..., \eta_n).$

Theorem: *The estimator η^0 defined in (14.19) is optimal.*

Proof: Applying the formula for the complete mathematical expectation, we find that

---

[7]The probabilities $P_\xi(x|\eta_1, ..., \eta_n)$ are sometimes called a posteriori (after considering $\eta_1, ..., \eta_n$).

$$P\{\eta \neq \xi\} = MP\{\eta \neq \xi \mid \eta_1, ..., \eta_n\}$$

$$= M(1 - P\{\xi = \eta|\eta_1, ..., \eta_n\})$$

$$= 1 - MP\{\xi = \eta|\eta_1, ..., \eta_n\}$$

$$\geqslant 1 - MP\{\xi = \eta^0|\eta_1, ..., \eta_n\}$$

$$= P\{\eta^0 \neq \xi\},$$

because, according to (14.19),

$$P\{\xi = \eta^0|\eta_1, ..., \eta_n\} \geqslant P\{\xi = \eta|\eta_1, ..., \eta_n\}.$$

Thus, the theorem is proved.   □

Section 15
A Filtration Problem
Kalman-Bucy Filter

We consider again a random process $\xi(t)$, $t \geqslant t_0$, with stochastic differential

(15.1) $d\xi(t) = \theta(t)dt + d\eta(t)$

and initial value $\xi(t_0) = 0$ where, contrary to (14.1), $\theta(t)$ is a random function (continuous in quadratic mean). We may interpret $\theta(t)$, $t \geqslant t_0$, as a "signal", which has to be filtered out from the superposed (random) noise, which is characterized in (15.1) by the standard Wiener process $\eta(t)$, $t \geqslant t_0$. We suppose that $\eta(t)$, $t \geqslant t_0$, does not depend on $\theta(t)$, $t \geqslant t_0$.

We have to estimate $\theta(t)$, given $\xi(s)$, $t_0 \leqslant s \leqslant t$. We shall obtain a fundamental result for such random functions $\theta(t)$, $t \geqslant t_0$, which satisfy a linear stochastic differential equation

(15.2) $d\theta(t) = a(t)\theta(t)dt + d\eta_0(t)$

with initial condition $\theta(t_0) = 0$, where $a(t)$ is a non-random continuous function and $\eta_0(t)$, $t \geqslant t_0$, is a standard Wiener process, which does not depend on $\eta(t)$, $t \geqslant t_0$. Here, both stochastic differentials (15.1), (15.2) are defined with respect to the same σ-algebras B^t, $t \geqslant t_0$, where each σ-algebra represents the events from the past up to the corresponding time t.

We shall consider linear estimators for $\theta(t)$, represented as linear combinations of the "observed" variables $\xi(s)$, $t_0 \leqslant s \leqslant t$, or their limit in quadratic mean.

Obviously, each linear combination of the variables $\xi(s)$, $t_0 \leqslant s \leqslant t$ (for example, the linear combination of the variables $\xi(t_0)$, $\xi(t_1)$, ..., $\xi(t_n)$, where $t_0 < t_1 < ... < t_n$) can be written as a stochastic integral

$$\eta = \sum_{k=1}^{n} c_k[\xi(t_k) - \xi(t_{k-1})] = \int_{t_0}^{t} c(s)d\xi(s)$$

with corresponding piecewise constant functions

$$c(s) = c_k, \quad t_{k-1} < s \leqslant t_k.$$

The limits of such linear combinations are estimators, represented as stochastic integrals

(15.3) $\qquad \eta = \int_{t_0}^{t} c(s)d\xi(s) = \int_{t_0}^{t} c(s)\theta(s)ds + \int_{t_0}^{t} c(s)d\eta(s)$

with an arbitrary function $c(s)$, $t_0 \leqslant s \leqslant t$, for which the last two integrals exist. Their sum will serve us as the definition of the stochastic integral with respect to $d\xi(s)$, which we shall use from now on. (For example, the linear estimator (15.3) is defined for any continuous function $c(s)$, $t_0 \leqslant s \leqslant t$.)

We note, that the orthogonality of any variables

$$\eta_1 = \int_{t_0}^{t} c_1(s)\theta(s)ds, \quad \eta_2 = \int_{t_0}^{t} c_2(s)d\eta(s)$$

follows from the condition of independence of the random process $\theta(t)$, $t \geqslant t_0$, and the standard Wiener process $\eta(t)$, $t \geqslant t_0$, in the Hilbert space H of random variables ξ, $M|\xi|^2 < \infty$, since η_1, η_2 are the limits in quadratic mean of the corresponding "integral sums"

$$\eta_{1n} = \sum_{k=1}^{n} c_{1k}\theta(t_{k-1})[t_k - t_{k-1}],$$

$$\eta_{2n} = \sum_{k=1}^{n} c_{2k}[\eta(t_k) - \eta(t_{k-1})]$$

and, obviously,

$$(\eta_1, \eta_2) = M\eta_1\eta_2 = \lim M\eta_{1n}\eta_{2n} = 0.$$

(For simplicity, from now on, we consider only real variables.)

So, the linear estimator (15.3) is the sum of the orthogonal variables

$$\eta_1 = \int_{t_0}^{t} c(s)\theta(s)ds, \quad \eta_2 = \int_{t_0}^{t} c(s)d\eta(s),$$

where

$$\|\eta_2\|^2 = M|\eta_2|^2 = \int_{t_0}^{t} |c(s)|^2 ds.$$

Problem: Show that the linear estimator (15.3) is defined for an arbitrary measurable function, satisfying the condition

(15.4) $\qquad \int_{t_0}^{t} |c(s)|^2 ds < \infty .$

Hint: Apply the fact that the random function $\theta(s)$, $t_0 \leqslant s \leqslant t$, is continuous in quadratic mean, and that the non-random function $c(s)$ may be represented as the limit of the piecewise constant functions

$$c_n(s) = c_{nk}, \quad t_{k-1} < s \leqslant t_k; \quad t_0 < t_1 < \cdots < t_n = t,$$

for which

$$\int_{t_0}^t |c(s) - c_n(s)|^2 ds \to 0.$$

Problem: Show that the linear estimators are given by the stochastic integral (15.3) with corresponding function $c(s)$, satisfying condition (15.4).

Let us denote by H the set of all variables $\eta \in \mathbf{H}$, described in (15.3), (15.4). Among all estimators $\eta \in H$, we shall look for the optimal estimator

$$(15.5) \qquad \hat{\theta}(t) = \int_{t_0}^t c(t,s) d\xi(s)$$

with weight function $c(t,s)$, $t \geqslant s$, which is the projection of the variable $\theta(t)$ on the linear subspace $H \subset \mathbf{H}$ in our Hilbert space $\mathbf{H}$ of random variables ξ.

The projection $\hat{\theta}(t)$ of the variable $\theta(t)$ on H is uniquely defined by the orthogonality condition

$$(15.6) \qquad \mathbf{M}[\theta(t) - \hat{\theta}(t)]\eta = 0$$

for a system of variables η of the form (15.3), which is complete in H. If we express this condition as a condition directly for the weight function $c(t,s)$ in (15.5), we can draw the following conclusion:

Lemma. *Suppose that the function $c(t,s)$ is continuous for all parameters $t \geqslant s \geqslant t_0$ and satisfies the integral equation*

$$(15.7) \qquad c(t,s) = B(t,s) - \int_{t_0}^t c(t,u)B(u,s)du, \quad t \geqslant s,$$

where $B(t,s) = \mathbf{M}\theta(t)\theta(s)$, $t,s \geqslant t_0$, is the correlation function of the random process $\theta(t)$, $t \geqslant t_0$. Then $c(t,s)$ is the weight function of the optimal estimator (15.5), and

$$(15.8) \qquad c(t,t) = \mathbf{M}[\theta(t) - \hat{\theta}(t)]^2$$

is the corresponding error in quadratic mean

Proof: In fact, if equation (15.7) holds, then we have the orthogonality condition (15.6), since we have for arbitrary variables

η of the form (15.3) that

$$\mathbf{M}[\theta(t) - \hat{\theta}(t)]\eta$$

$$= \mathbf{M}\left[\theta(t) - \int_{t_0}^t c(t,u)\theta(u)du - \int_{t_0}^t c(t,u)d\eta(u)\right]$$

$$\times \left[\int_{t_0}^t c(s)\theta(s)ds + \int_{t_0}^t c(s)d\eta(s)\right]$$

$$= \int_{t_0}^t c(s)\left[B(t,s) - \int_{t_0}^t c(t,u)B(u,s)du - c(t,s)\right]ds = 0.$$

Now, at this point we apply the general formulas (7.12) and (7.13) to compute the mathematical expectation. Furthermore

$$\mathbf{M}[\theta(t) - \hat{\theta}(t)]^2 = \mathbf{M}\theta(t)[\theta(t) - \hat{\theta}(t)] - \mathbf{M}\hat{\theta}(t)[\theta(t) - \hat{\theta}(t)],$$

where

$$\mathbf{M}\theta(t)[\theta(t) - \hat{\theta}(t)]$$

$$= \mathbf{M}\theta(t)^2 - \mathbf{M}\left[\theta(t) \cdot \int_{t_0}^t c(t,u)\theta(u)du\right]$$

$$- \mathbf{M}\left[\theta(t) \cdot \int_{t_0}^t c(t,u)d\eta(u)\right]$$

$$= B(t,t) - \int_{t_0}^t c(t,u)B(u,t)du = c(t,t)$$

according to (15.7) with $s = t$; and

$$\mathbf{M}\hat{\theta}(t)[\theta(t) - \hat{\theta}(t)] = 0$$

by the orthogonality condition (15.6) for $\eta = \hat{\theta}(t)$. Hence, the lemma is proved.   □

Now, let $\theta(t)$, $t \geqslant t_0$, be a random process with stochastic differential (15.2); with the initial condition $\theta(t_0) = 0$, it looks like

(15.9) $\theta(t) = \int_{t_0}^t \omega_0(t,s)d\eta_0(s), \quad t \geqslant t_0 ,$

where $\omega_0(t,s)$, $t \geqslant s$, is the solution of the differential equation

$$\frac{d}{dt} \omega_0(t,s) = a(t)\omega_0(t,s), \quad t > s,$$

with initial condition $\omega_0(s,s) = 1$. We have

$$\theta(t) = \omega_0(t,s)\theta(s) + \int_s^t \omega_0(t,u)d\eta_0(u), \quad t > s,$$

where the second term of the sum is a variable, which is orthogonal to $\theta(s)$, and

$$B(t,s) = M\theta(t)\theta(s) = \omega_0(t,s)B(s,s), \quad t \geqslant s.$$

It is immediately obvious that the correlation function $B(t,s)$ satisfies the differential equation

$$(15.10) \qquad \frac{d}{dt} B(t,s) = a(t)B(t,s), \quad t > s.$$

We try to find a function $c(t,s)$ which is continuous with the derivative $\frac{d}{dt} c(t,s)$ for all parameters $t \geqslant s \geqslant t_0$ and for which the integral equation (15.7) holds.

By differentiating equation (15.7) with respect to t and with condition (15.10), we obtain for such a function:

$$\frac{d}{dt} c(t,s) = a(t)B(t,s) - c(t,t)B(t,s) - \int_{t_0}^t \frac{d}{dt} c(t,u)B(u,s)du.$$

If we introduce for fixed s the function $x(t)$ defined by

$$\frac{d}{dt} c(t,s) = x(t)c(t,s), \quad t > s,$$

then we can derive with (15.7)

$$\dot{x}(t)B(t,s) = x(t)\left[c(t,s) + \int_{t_0}^t c(t,u)B(u,s)du\right]$$

$$= [a(t) - c(t,t)]B(t,s).$$

Obviously, $x(t) = a(t) - b(t)$, where

$$b(t) = c(t,t).$$

Having obtained this unexpected result, it is natural to look for the function $c(t,s)$, $t \geqslant s$, as the solution of a differential equation of the form

$$(15.11) \qquad \frac{d}{dt} c(t,s) = [a(t) - b(t)]c(t,s), \quad t > s, \quad c(s,s) = b(s), \quad s \geqslant t_0.$$

We take a continuous function $b(t)$, $t \geqslant t_0$. Then the solution $c(t,s)$ of the linear differential equation (15.11) and the derivative $(d/dt)c(t,s)$ are continuous for all parameters $t \geqslant s \geqslant t_0$. If we take this solution $c(t,s)$ as a weight function, defining the random function $\hat{\theta}(t)$, $t \geqslant t_0$, by formula (15.5), then it is possible to express the stochastic differential $d\hat{\theta}(t)$ in a form similar to (9.19):

$$d\hat{\theta}(t) = \left[\int_{t_0}^t \frac{d}{dt} c(t,s)d\xi(s)\right]dt + c(t,t)d\xi(t)$$

or, with (15.11), in the form

$$d\hat{\theta}(t) = [a(t) - b(t)]\hat{\theta}(t)dt + b(t)[\theta(t)dt + d\eta(t)].$$

Because a stochastic differential $d\theta(t)$ of the form (15.2) exists, we obtain the following expression for the stochastic differential of the difference $\Delta(t) = \theta(t) - \hat{\theta}(t)$:

$$d\Delta(t) = d\theta(t) - d\hat{\theta}(t)$$

(15.12)

$$= [a(t) - b(t)]\Delta(t)dt + [d\eta_0(t) + b(t)d\eta(t)].$$

We can express the stochastic measure $d\eta_0(t) + b(t)d\eta(t)$ by the standard Wiener process $\zeta(t)$, $t \geqslant t_0$, which is defined by

$$\zeta(t) = \int_{t_0}^{t} \frac{1}{\sqrt{1 + b(s)^2}} d\eta_0(s) + \int_{t_0}^{t} \frac{b(s)}{\sqrt{1 + b(s)^2}} d\eta(s)$$

and setting

$$d\eta_0(t) + b(t)d\eta(t) = \overline{\sqrt{1 + b(t)^2}} \, d\zeta(t),$$

we obtain

(15.13) $$d\Delta(t) = [a(t) - b(t)]\Delta(t)dt + \overline{\sqrt{1 + b(t)^2}} \, d\zeta(t).$$

Obviously, the random function $\Delta(t)$, $t \geqslant t_0$, is the solution of a first order linear stochastic differential equation with initial condition $\Delta(t_0) = 0$. We have seen that we can write it in the form

$$\Delta(t) = \int_{t_0}^{t} \omega(t,s) \overline{\sqrt{1 + b(s)^2}} \, d\zeta(s)$$

(15.14)

$$= \int_{t_0}^{t} \omega(t,s)d\eta_0(s) + \int_{t_0}^{t} \omega(t,s)b(s)d\eta(s),$$

where the weight function $\omega(t,s)$, $t \geqslant s$, is the solution of the ordinary differential equation

$$\frac{d}{dt} \omega(t,s) = [a(t) - b(t)]\omega(t,s), \quad t > s,$$

$$\omega(s,s) = 1$$

(cf. (12.10)). If we compare this equation to (15.11), we see that

(15.15) $$c(t,s) = \omega(t,s)b(s), \quad t \geqslant s.$$

If the weight function $c(t,s)$ defines the optimal estimator (15.5), then the corresponding function $b(t) = c(t,t)$ is

(15.16) $$b(t) = M\Delta(t)^2, \quad t \geqslant t_0$$

(cf. (15.8)), and if $\Delta(t)$ is the solution of the linear stochastic differential equation (15.13) with initial value $\Delta(t_0) = 0$, then the function $b(t)$ is the solution of the corresponding ordinary differential equation

(15.17)
$$\frac{d}{dt} b(t) = 2a(t)b(t) - b(t)^2 + 1, \quad t > t_0,$$

$$b(t_0) = 0$$

(cf. the general formula (12.15)). Equation (15.17) is the well-known *Riccati equation*.

Now we take the function $b(t)$, $t \geqslant t_0$, which is the solution of this equation and we consider the random function $\Delta(t)$, $t \geqslant t_0$, which is the solution of the linear stochastic differential equation (15.12) - (15.13) with initial condition $\Delta(t_0) = 0$. Equation (15.16) holds for $\Delta(t)$, since both parts of it fulfill the equation (15.17). We set

$$\hat{\theta}(t) = \theta(t) - \Delta(t), \quad t \geqslant t_0.$$

With (15.2), we can express the stochastic differential $d\hat{\theta}(t) = d\theta(t) - d\Delta(t)$ in the form of the equation

(15.18) $$d\hat{\theta}(t) = [a(t) - b(t)]\hat{\theta}(t)dt + b(t)d\xi(t), \quad t > t_0.$$

(Recall the calculation (15.12).) The solution of this equation for $\hat{\theta}(t_0) = 0$ may be written in integral form:

(15.19)
$$\hat{\theta}(t) = \int_{t_0}^{t} c(t,s)d\xi(s)$$

$$= \int_{t_0}^{t} c(t,s)\theta(s)ds + \int_{t_0}^{t} c(t,s)d\eta(s)$$

with the weight function of (15.11), (15.15). (Cf. (12.10).)

We show that this weight function $c(t,s) = \omega(t,s)b(s)$ satisfies the integral equation (15.7). If we apply the expression (15.9), (15.14) and the differential equation (15.17) for $b(t) = M\Delta(t)^2$, we derive easily that the function

$$f(t) = M\hat{\theta}(t)[\theta(t) - \hat{\theta}(t)] = M[\theta(t) - \Delta(t)]\Delta(t)$$

$$= \int_{t_0}^{t} \omega_0(t,s)\omega(t,s)ds - b(t)$$

satisfies the homogeneous differential equation

$$\frac{d}{dt} f(t) = [2a(t) - b(t)]f(t),$$

$$f(t_0) = 0,$$

and, hence, $f(t) \equiv 0$. Therefore, with formula (15.19) we get

$$b(t) = M\theta(t)[\theta(t) - \hat{\theta}(t)] - M\hat{\theta}(t)[\theta(t) - \hat{\theta}(t)]$$

$$= M\theta(t)[\theta(t) - \hat{\theta}(t)] = B(t,t) - \int_{t_0}^{t} c(t,s)B(s,t)ds, \quad t \geqslant t_0.$$

Applying this expression for $b(t)$ and the equations (15.10), (15.11) for the function $c_0(t,s)$ defined by

$$c_0(t,s) = c(t,s) + \int_{t_0}^{t} c(t,u)B(u,s)du - B(t,s), \quad t \geqslant s,$$

we obtain the homogeneous equation

$$\frac{d}{dt}c_0(t,s) = [a(t) - b(t)]c_0(t,s), \quad t > s,$$

$$c_0(s,s) = 0,$$

and, hence, $c_0(t,s) \equiv 0$; we obtained the integral equation (15.7) for our weight function $c(t,s)$.

Let us present our final result as a theorem:

Theorem. *The optimal estimator $\hat{\theta}(t)$ for $\theta(t)$ is given by the stochastic integral (15.5) with weight function $c(t,s)$, $t \geqslant s$, which is together with the function $b(t) = M[\theta(t) - \hat{\theta}(t)]^2$ the solution of the system of differential equations (15.11), (15.17).*

The method of finding an optimal estimator that we applied here is usually called *Kalman-Bucy method*. It amounts to solving the stochastic differential equation (15.18).

Appendix
Basic Concepts of Probability Theory

On the following pages we want to outline the general framework of probability theory. Probability theory assumes as given a set Ω of *elementary outcomes* ω, the so-called elementary events, and a σ-algebra A of sets $A \subseteq \Omega$, the so-called *events*. Furthermore, for all $A \in A$, a probability $P(A)$ is defined. P is a non-negative measure on the σ-algebra of events A satisfying $P(\Omega) = 1$. The *event* A *consists of the elementary outcomes* $\omega \in A$, so that the event A and the complementary event $A^c = \Omega \backslash A$ are mutually exclusive.

Recall that a system of sets is called a *σ-algebra* if it is invariant with respect to taking unions, respectively intersections, respectively complements of a countable number of sets. A *probability measure* P on the σ-algebra A is a function $P: A \longmapsto [0,1]$, which is σ-additive in the sense that

$$P\left[\underset{k}{\cup} A_k\right] = \underset{k}{\Sigma} P(A_k)$$

for disjoint events A_k, $k = 1,2, ...,$ and satisfies $P(\Omega) = 1$.

In applied probabilistic models, we are usually given the probabilities of some relatively "simple" events, and the task is to determine the probabilities of some "complex" events. If the "simple" events form a semi-ring[8], we can derive the probabilities of any

---

[8]Here we follow the terminology of the book: A. N. Kolmogorov, S. Fomin, *Introductory Real Analysis*, Russian: 5th ed., Moscow, Nauka, 1981; English: New York, Dover Publications, 1975.

A system A of sets is called a semi-ring, if it satisfies the following conditions:

(i) $A, A_1 \in A$ implies $A \cap A_1 \in A$
(ii)   if $A, A_1 \in A$ and $A_1 \subseteq A$, then there exists a finite number of sets $A_2, ..., A_n \in A$ such that $A \backslash A_1 = A_2 \cup A_3 \cup ... \cup A_n$

event A, which is generated by taking a countable number of unions, intersections or complements of "simple" events, in the following way:

$$P(A) = \inf_{k} \sum P(A_k),$$

where inf is taken over all "simple" event A_k, whose union contains A (we suppose that the probability measure defined on the initial semi-ring of events is σ-additive).

As an example for a "complex" event, which is expressed in terms of events A_1, A_2 we can take

$$A = \bigcap_{n=1}^{\infty} \left[\bigcup_{k=n}^{\infty} A_1 \right];$$

the event A is a realization of infinitely many of the events A_1, A_2. If $\sum_{k=1}^{\infty} P(A_k) < \infty$, then $P(A) = 0$, i.e., with probability 1 only a finite number of events is realized. (This result is well-known as *Borel-Cantelli-Lemma*.)

Ordinarily, in an actual probabilistic model, certain dependence structures hold. In such cases, the above-mentioned procedure for computing probabilities of complex events is not very helpful.

Next, we recall the notion of *independence*. Suppose we have a probability space (Ω, A, P); two sets $A_1, A_2 \in A$ are defined as being independent, if

$$P(A_1 \cdot A_2) = P(A_1) \cdot P(A_2).$$

(Sometimes, we write $A_1 \cdot A_2$ instead of $A_1 \cap A_2$.) The events A_1, ..., A_n are called *independent*, if

$$P(A_{i_1} \dots A_{i_m}) = P(A_{i_1}) \dots P(A_{i_m})$$

for all subsets $\{i_1, ..., i_m\} \subseteq \{i, ..., n\}$.

The degree of dependence of an event A on an event B is given by the *conditional probability*, which is defined as

$$P(A|B) = \frac{P(AB)}{P(B)} \quad \text{for} \quad P(B) > 0.$$

Let $B \in A$ be given with $P(B) > 0$. Then the probability under condition B is the *conditional probability measure* on it defined by

$$A \longmapsto P(A|B), \quad A \in A.$$

---

[8](iii) there exists a sequence B_k, $k = 0,1,2, \dots$ such that

$$A = \bigcup_{k \geqslant 0} B_k.$$

A further fundamental concept in probability theory is the notion of a *random variable*. A random variable is defined as a measurable function

$$\xi = \xi(\omega), \quad \omega \in \Omega ,$$

on the probability space (Ω, A, P); usually, we think of ξ as a real function defined for the elementary outcomes $\omega \in \Omega$; we consider also complex random variables of the form $\xi = \xi_1 + i\xi_2$ with real components ξ_1, ξ_2, multi-dimensional variables $\xi = (\xi_1, ..., \xi_n) \in \mathbb{R}^n$, the components of which are real random variables and so on.

Let $\xi_1, ..., \xi_n$ be random variables with values in $\mathbb{R}^1$; then the random vector $\xi = (\xi_1, ..., \xi_n)$ has values in $\mathbb{R}^n$. Its *probability distribution* on the space of measurable sets $B \subseteq \mathbb{R}^n$ is defined by

$$P(\xi \in B\} = P_\xi(B).$$

P_ξ is determined by its value on the semi-ring of sets $B_1 \times ... \times B_n$, where the B_i are measurable subsets of $\mathbb{R}^1$:

$$P\{\xi_1 \in B_1, ..., \xi_n \in B_n\} = P_\xi(B_1 \times ... \times B_n),$$

$$B_1, ..., B_n \in \mathbb{R}^1.$$

If we have a finite number of random variables $\xi_1, ..., \xi_n$, it is convenient to suppose that they are defined on the space $\Omega = \mathbb{R}^n$ of elemenary outcomes $\omega = x$ with probability measure $P(d\omega) = P_\xi(dx)$ setting $\xi_1(x) = x_1, ..., \xi_n(x) = x_n$ for $x = (x_1, ..., x_n)$, $x \in \mathbb{R}^n$.

The variables $\xi_1, ..., \xi_n$ are called *independent*, if their distribution in the n-fold product space $\mathbb{R}^n$ is the product of the distributions in $\mathbb{R}^1$ of the variables ξ_k, i.e.,

$$P\{\xi_1 \in B_1, ..., \xi_n \in B_n\} = P\{\xi_1 \in B_1\} \cdot ... \cdot P\{\xi_n \in B_n\}.$$

If we are dealing with random variables and events jointly, it is convenient to identify the events $A \in \Omega$ with their indicator function

$$1_A(\omega) = \begin{cases} 1, & \omega \in A, \\ 0, & \omega \notin A . \end{cases}$$

Let $\xi(t)$, $t \in T$, be an arbitrary collection of random variables (the index $t \in T$ is arbitrary). By A_ξ we denote the smallest σ-algebra of events with respect to which all $\xi(t)$ are measurable.

$$\xi(t) = \xi(\omega, t), \quad \omega \in \Omega$$

A random variable $\xi = \xi(\omega)$, which is measurable with respect to A_ξ, is called a function of the variable $\xi(t)$, $t \in T$. Arbitrary systems of random variables $\xi(t)$, $t \in T$, and $\eta(s)$, $s \in S$, are *independent*, if the following conditions hold: Any two variables ξ and η are independent, and ξ is measurable with respect to A_ξ and η is measurable with respect to A_η.

The relationship between the random variables $\xi \in \mathbb{R}^n$ and $\eta \in \mathbb{R}^m$ can be expressed by the *conditional probability distribution* of the variable ξ with respect to η. The conditional probability distribution is defined as the measure $P_\xi(dx|y)$ in $\mathbb{R}^n$, which satisfies the following properties:

(i) for fixed $B \subset \mathbb{R}^n$, the function $y \longmapsto P_\xi(B|y)$ is measurable in $\mathbb{R}^m$.

(ii) $P\{\xi \in B, \eta \in \Gamma\} = \int_\Gamma P_\xi(B|y)P_\eta(dy)$, $B \subseteq \mathbb{R}^n$, $\Gamma \subseteq \mathbb{R}^m$.

The probabilities $P_\xi(B|y) = P\{\xi \in B | \eta = y\}$ are called *conditional* under the condition $\eta = y$. As a typical example we take the conditional probability distribution of the variable $\xi = \varphi(\eta,\eta')$, which is the function of two independent random variables η and η'. The resulting $P_\xi(dx|y)$ is the distribution of the random variable $\varphi(y,\eta')$ for fixed y, which follows immediately from the formula for multiple integration with respect to the product of the measures $P_\eta(dy) \times P_{\eta'}(dy')$:

$$P\{\varphi(\eta,\eta') \in B, \eta \in \Gamma\} = \int_\Gamma \left[\int_{\{y':\varphi(y,y') \in B\}} P_{\eta'}(dy') \right] P_\eta(dy)$$

$$= \int_\Gamma P\{\varphi(y,\eta') \in B\} P_\eta(dy).$$

Concerning the notion of conditional distribution, let us make the following remark: independence of the random variables ξ and η implies that the conditional distribution $P_\xi(dx|y)$ does not depend on y and is given by

$$P_\xi(dx|y) = P_\xi(dx).$$

The random variable ξ is said to be *conditionally independent* of ζ, given a variable η, if the conditional distribution of the variable ξ, given the random variable (η,ζ), does not depend on ζ, more precisely, if for any condition $\eta = y$, $\zeta = z$

$$P_\xi(dx|y) = P_\xi(dx|y, z).$$

For example, if $\xi = \varphi(\eta,\eta')$ is a function on η and η', where η' does not depend on (η,ζ), we can infer that ξ is conditionally independent of ζ for a given η.

Given a random variable ξ, we define its *mathematical expectation*, which is sometimes also called *mean value*, by the integral

$$\mathbf{M}\xi = \int_\Omega \xi(\omega)\mathbf{P}(d\omega).$$

Given a sequence of independent, identically distributed random variables

$$\xi = \xi_1, \xi_2, \dots$$

with mathematical expectation $\mathbf{M}\xi$, the *strong law of large numbers* holds, i.e.,

$$\lim_{n\to\infty} \frac{1}{n} \sum_{k=1}^{n} \xi_k = \mathbf{M}\xi .$$

with probability 1.

The following inequality holds for the mathematical expectation:

$$\mathbf{M}|\xi\eta| \leqslant (\mathbf{M}\,|\xi|^p)^{1/p}(\mathbf{M}\,|\eta|^q)^{1/q},$$

where $p,q > 0$, $1/p + 1/q = 1$. Furthermore, for independent random variables ξ, η

$$\mathbf{M}\xi\eta = \mathbf{M}\xi \cdot \mathbf{M}\eta.$$

For the random variable $\varphi(\xi)$, where $\xi = (\xi_1, \dots, \xi_n)$ and $\varphi(x)$ is a function of $x \in \mathbb{R}^n$ that is integrable with respect to $\mathbf{P}_\xi(dx)$, the following formula holds:

$$\mathbf{M}\varphi(\xi) = \int_{\mathbb{R}^n} \varphi(x)\mathbf{P}_\xi(dx).$$

The *variance*

$$\mathbf{D}\xi = \mathbf{M}(\xi - \mathbf{M}\xi)^2$$

characterizes the deviation in quadratic mean of the random variable ξ from its expectation $\mathbf{M}\xi$. The *correlation* of (real) random variables ξ,η is defined as

$$\mathbf{M}(\xi - \mathbf{M}\xi)(\eta - \mathbf{M}\eta).$$

The set of all random variables ξ, $\mathbf{M}|\xi|^2 < \infty$ on the probability space $(\Omega, \mathcal{A}, \mathbf{P})$ forms a Hilbert space $\mathbf{H}$ with the scalar product

$$(\xi,\eta) = \mathbf{M}\xi\overline{\eta} = \int_\Omega \xi(\omega)\overline{\eta(\omega)}\,\mathbf{P}(d\omega), \quad \xi,\eta \in \mathbf{H},$$

where $\overline{\eta}$ denotes the conjugate variable (in $\mathbf{H}$, we have to identify variables that are equal with probability 1, i.e., if they are equal for almost every $\omega \in \Omega$). The Hilbert space thus defined of measurable functions $\xi = \xi(\omega)$ on Ω, that are integrable in quadratic mean, with norm

$$\| \xi \| = (\mathbf{M}|\xi|^2)^{1/2} = \left[\int_\Omega |\xi(\omega)|^2 \mathbf{P}(d\omega) \right]^{1/2}$$

is well-known as L^2-space. The norm

$$\| \xi - \eta \| = (\mathbf{M}|\xi - \eta|^2)^{1/2}$$

is the so-called *distance in quadratic mean* between the variables $\xi, \eta \in$ H. Having defined a distance measure, we can define the *convergence in quadratic mean*, $\xi_n \to \xi$. Set $\xi = \lim_{n\to\infty}\xi_n$, then ξ_n is said to converge against ξ if

$$\| \xi - \xi_n \| \to 0.$$

ξ is called the *limit in quadratic mean*. H is a complete Hilbert space, i.e., each fundamental sequence $\xi_n \in$ H,

$$\| \xi_n - \xi_m \| \to 0, \quad n,m \to \infty$$

has a limit (in quadratic mean) $\xi = \lim_{n\to\infty}\xi_n$ in H.

The following simple inequality holds as a consequence of the well-known *Chebyshev-inequality*:

$$\mathbf{P}\{|\xi| \geqslant \epsilon\} \leqslant \frac{1}{\epsilon} \| \xi \|.$$

The probability distribution of a variable $\xi \in \mathbb{R}^n$ may be expressed by the conditional probability $\mathbf{P}_\xi(dx|y)$ with respect to a random variable $\eta \in \mathbb{R}^m$, using the *formula of complete probability*:

$$\mathbf{P}_\xi(B) = \mathbf{M}\mathbf{P}_\xi(B|\eta), \quad B \subseteq \mathbb{R}^n.$$

The analogue for $\xi \in \mathbb{R}^1$ is the *formula of complete mathematical expectation*

$$\mathbf{M}\xi = \mathbf{M}[\mathbf{M}(\xi|\eta)],$$

where

$$\mathbf{M}(\xi|y) = \int_{\mathbb{R}^1} \mathbf{P}_\xi(dx|y), \quad y \in \mathbb{R}^m,$$

the so-called *conditional mathematical expectation* of the variable ξ with respect to η. We may interpret the value $\mathbf{M}(\xi|y) = \mathbf{M}(\xi|\eta = y)$ as the mathematical expectation of the variable ξ under the condition $\eta = y$.

One of the most important distributions in probability theory is the *Gaussian distribution* $\mathbf{P}(dx)$, which is sometimes also called *normal distribution*. On the real line $\mathbb{R}^1$ it is given by the density

$$p(x) = \frac{1}{\sqrt{2\pi}\sigma} e^{-\frac{1}{2}\frac{(x-a)^2}{\sigma^2}}, \quad x \in \mathbb{R}^1.$$

In the n-fold product space $\mathbb{R}^n$ it may be expressed, for example, by the *characteristic function*

$$\varphi(u_1, ..., u_n) = \int_{\mathbb{R}^n} e^{i\sum_{k=1}^n u_k x_k} P(dx), \quad u_1, ..., u_n \in \mathbb{R}^1,$$

which, for the Gaussian distribution $P(dx)$, $x = (x_1, ..., x_n)$ takes the form

$$\varphi(u_1, ..., u_n) = \exp\left\{ i \sum_{k=1}^n A_k u_k - \frac{1}{2} \sum_{k,j=1}^n B_{kj} u_k u_j \right\},$$

where the linear term of the variables $u_1, ..., u_n$ is arbitrary, and the quadratic term is non-negative.

The *central limit theorem* attributes special importance to the normal distribution. According to this theorem, random variables that are a sum of a large number of weakly dependent terms are distributed approximately normally. *In the case of a sequence of sums* $S_n = \sum_{k=1}^n \xi_k$ *of independent, identically distributed variables with expectation a and variance* σ^2, *we get the most simple form of the central limit theorem as follows*:

$$\lim_{n \to \infty} \mathbf{P}\left\{ \frac{S_n - na}{\sigma \sqrt{n}} \leqslant x \right\} = \frac{1}{\sqrt{2\pi}} \int_{-\infty}^x e^{-x^2/2} dx.$$

Random variables $\xi = (\xi_1, ..., \xi_n)$ with *Gaussian distribution* $P_\xi(dx) = P(dx)$ are called Gaussian. The corresponding parameters A_k and B_{kj} can be interpreted in probabilistic terms in a simple way. Given the expectation and the correlation of the variables $\xi_1, ..., \xi_n$, it can be derived

$$A_k = \mathbf{M}\xi_k, \quad B_{kj} = \mathbf{M}(\xi_k - A_k)(\xi_j - A_j), \quad k,j = 1, ..., n.$$

If the $(n \times n)$ correlation matrix $\{B_{kj}\}$ is nondegenerated with determinant σ^2, the Gaussian distribution $P(dx)$ has the density

$$p(x) = \frac{1}{(2\pi)^{n/2}\sigma} \exp\left\{ -\frac{1}{2} \sum_{k,j=1}^n b_{kj}(x_k - A_k)(x_j - A_j) \right\},$$

where $\{b_{kj}\} = \{B_{kj}\}^{-1}$ is the inverse matrix.

The density $p(x)$, $x \in \mathbb{R}^n$, of an arbitrary probability distribution is called *probability density*, and we write:

$$\mathbf{P}(B) = \int_B p(x)dx, \quad B \subseteq \mathbb{R}^n.$$

The above mentioned central limit theorem is an example for a class of limit theorems in probability theory, which gives conditions for the *weak convergence* of distributions to some "standard" distribution.

A sequence of distributions $P_n(dx)$, $-\infty < x < \infty$, of random variables ξ_n is said to *converge weakly* towards the distribution $P(dx)$ of the random variable ξ (in mathematical notation: $P_n \rightarrow P$), if

$$\int_{x'}^{x''} P_n(dx) = P\{x' < \xi_n \leqslant x''\} \rightarrow P\{x' < \xi \leqslant x''\} = \int_{x'}^{x''} P(dx)$$

for all x', x'' which satisfy

$$P\{\xi = x'\} = P\{\xi = x''\} = 0.$$

If $\xi_n \rightarrow \xi$ is convergent in quadratic mean, then the corresponding probability distributions P_n converge weakly to P, i.e.

$$\int_{x'}^{x''} \varphi(x)P_n(dx) \rightarrow \int_{x'}^{x''} \varphi(x)P(dx).$$

for any continuous function $\varphi(x)$ and the above mentioned x', x''.

A *random process* on the set $T \subseteq \mathbf{R}^1$ is defined as a family of random variables $\xi(t)$, depending on the real parameter $t \in T$. In general, t is interpreted as time parameter. We use the term of a random process $\xi(t)$, $t \in T$, synonymously with the term *random function* of $t \in T$; its values are the random variables $\xi(t)$. Interpreted in practical terms, we can say that the random process $\xi(t)$, $t \in T$, describes the evolution of some "system" whose "state" at time t is $\xi(t)$. Thus, we call $\xi(t)$ the *state* of the random process at time t.

Of course, if we deal with the random process $\xi(t)$, $t \in T$, we have to say that its values are defined on some probability space (Ω, A, P) i.e., $\xi(t) = \xi(\omega,t)$, $\omega \in \Omega$. For fixed $\omega \in \Omega$, we speak of the *trajectory* (or the *sample path*, or the *realization*) $\xi(\omega,\cdot) = \xi(\omega,t)$, $t \in T$, of this random process $\xi(t)$.

To be definite, from now on, we shall consider real variables $\xi(t)$, $t \in T$.

Let X be some space of real functions $x = x(t)$, $t \in T$, containing all trajectories of the random process $\xi(t)$, $t \in T$. The application $\omega \rightarrow x = \xi(\omega,\cdot)$ allows to introduce on X the σ-algebra B, which is generated by all sets $B \subseteq X$ in such a way that $\{\xi(\omega,\cdot) \in B\} \in A$ and with probability measure $P_\xi(B) = P\{\xi(\omega,\cdot) \in B\}$. Defining the random variables $\xi(t)$ on the probability space (X, B, P), with $\xi(x,t) = x(t)$, $x \in X$, $t \in T$, it becomes obvious that for such a random process $\xi(t)$, $t \in T$ (which is called *canonically represented*) the probability distribution of any of its values $\xi(t_1)$, ..., $\xi(t_n)$ is the same as for the original process $\xi(t)$, $t \in T$. Here, we take as elementary outcomes of the new (canonically represented) random process its *trajectories* $x = \xi(x,t)$, $t \in T$ (we may say, that each of them describes one of the possible realizations of the observed process).

Given *finite distributions*

$$\mathbf{P}_{t_1,\dots,t_n}(B_1 \times \dots \times B_n) \qquad B_1, \dots, B_n \subseteq \mathbb{R}^1,$$

naturally the question arises, whether such a family of random variables $\xi(t)$, $t \in T$, exists such that

$$P\{\xi(t_1) \in B_1, \dots, \xi(t_n) \in B_n\} = \mathbf{P}_{t_1,\dots,t_n}(B_1 \times \dots \times B_n)$$

$$B_1, \dots, B_n \in \mathbb{R}^1.$$

More precisely, is it possible to realize such a family on some probability space $(\Omega, \mathsf{A}, \mathbf{P})$ or on some function space $(X, \mathsf{B}, \mathbf{P}_\xi)$? The answer is the fundamental *Kolmogorov*[9] *theorem*, which we shall formulate in the following way:

Suppose that a family of consistent probability distributions

$$\mathbf{P}_{t_1,\dots,t_n}(B_1 \times \dots \times B_n), \quad B_1, \dots, B_n \subseteq \mathbb{R}^1,$$

is given; then a random process $\xi(t)$, $t \in T$, *exists, which has these finite-dimensional distributions. The notion of consistency mentioned here, which has to be satisfied for finite dimensional distributions of an arbitrary random process, implies that for any* $t_1, \dots, t_n \in T$

$$\mathbf{P}_{t_{k_1},\dots,t_{k_n}}(B_{t_{k_1}} \times \dots \times B_{t_{k_n}}) = \mathbf{P}_{t_1,\dots,t_n}(B_1 \times \dots \times B_n)$$

for any fixed (t_k, B_k) *and that*

$$\mathbf{P}_{t_1,\dots,t_{n-1},t_n}(B_1 \times \dots \times B_{n-1} \times \mathbb{R}^1)$$

$$= \mathbf{P}_{t_1,\dots,t_{n-1}}(B_1 \times \dots \times B_{n-1}).$$

It turns out that we can always take as corresponding probability space $(\Omega, \mathsf{A}, \mathbf{P})$ the space $\Omega = X$ of all real functions $x = x(t)$, $t \in T$, with σ-algebra A, generated by the *semi-ring* A_0 of the *cylindric sets* $A \subseteq X$ of the form $A = \{x: x(t_1) \in B_1, \dots, x(t_n) \in B_n\}$ with arbitrary $t_1, \dots, t_n \in T$, $B_1, \dots, B_n \subseteq \mathbb{R}^1$ and $n = 1, 2, \dots$. The probability measure $\mathbf{P} = \mathbf{P}_\xi$, for which the canonically represented random process $\xi(t)$, $t \in T$, has the given finite dimensional distributions $\mathbf{P}_{t_1,\dots,t_n}$, is defined on the above mentioned cylindrical sets as

$$\mathbf{P}(A) = \mathbf{P}_{t_1,\dots,t_n}(B_1 \times \dots \times B_n)$$

and is extended to the whole σ-algebra A.

---

[9]Cf. A. N. Kolmogorov, Foundations of the theory of probability theory, Russian: 2nd ed., Moscow, Nauka, 1974; English: 2nd Engl. ed., New York, Chelsea Publishing Company, 1956.

We can express the consistency in terms of characteristic functions

$$\varphi_{t_1,\ldots,t_n}(u_1, \ldots, u_n)$$

$$= \int_{\mathbb{R}^n} e^{i\Sigma_{k=1}^n u_k x_k} P_{t_1,\ldots,t_n}(dx_1 \ldots dx_n), \quad u_1, \ldots, u_n \in \mathbb{R}^1.$$

In fact, the condition of consistency is equivalent to the condition that

$$\varphi_{t_{k_1},\ldots,t_{k_n}}(u_{k_1}, \ldots, u_{k_n}) = \varphi_{t_1,\ldots,t_n}(u_1, \ldots, u_n),$$

$$\varphi_{t_1,\ldots,t_{n-1},t_n}(u_1, \ldots, u_{n-1}, 0) = \varphi_{t_1,\ldots,t_{n-1}}(u_1, \ldots, u_{n-1}).$$

For example, let us analyze the case of the consistent family of Gaussian (normal) probability distributions $P_{t_1,\ldots,t_n}$ with characteristic functions

$$\varphi_{t_1,\ldots,t_n}(u_1, \ldots, u_n)$$

$$= \exp\left\{ i \sum_{k=1}^n A(t_k)u_k \ - \frac{1}{2} \sum_{k,j=1}^n B(t_k,t_j)u_k u_j \right\}.$$

where $A(t)$, $t \in T$, is an arbitrary real function, and $B(t,s)$, $t,s \in T$, is a real function, satisfying the following condition of positive definiteness:

$$\sum_{k,j=1}^n B(t_k,t_j)u_k u_j \geqslant 0$$

for any real $u_1, \ldots, u_n$ and $t_1, \ldots, t_n \in T$. If Gaussian random variables $\xi(t_1), \ldots, \xi(t_n)$ are given, then the meaning of A respectively B is

$$A(t_k) = M\xi(t_k),$$

$$B(t_k,t_j) = M[\xi(t_k) - A(t_k)][\xi(t_j) - A(t_j)], \quad k,j = 1, \ldots, n.$$

A random process $\xi(t)$, $t \in T$, whose finite dimensional distributions $P_{t_1,\ldots,t_n}$ are Gaussian, is called Gaussian. In general, we call the function $A(t) = M\xi(t)$, $t \in T$, the mean value (expectation) and the function

$$B(t,s) = M[\xi(t) - A(t)][\xi(s) - A(s)], \quad t,s \in T$$

the correlation function of the random process $\xi(t)$, $t \in T$.

Y. A. Rozanov

Markov Random Fields

Translated from the Russian by
C. M. Elson

1982. 1 figure. IX, 201 pages.
ISBN 3-540-90708-4

Contents: General Facts About Probability
Distributions. – Markov Random Fields. –
The Markov Property for Generalized
Random Functions. – Vector-Valued
Stationary Functions. – Notes. – Biblio-
graphy. – Index.

The investigations of random fields with
the Markov property has developed into
one of the most interesting research topics
of the last few years. This book summarizes
the field and many of the recent devel-
opments in it for scientists who wish to
utilize this theory for applications in physics
or engineering. An important factor in this
development is the fact that only recently
have whole classes of various random functions been found which possess the
Markov property.

Springer-Verlag
Berlin Heidelberg New York
London Paris Tokyo

D. Dacunha-Castelle, M. Duflo

Probability and Statistics I

Translated by D. McHale

1986. 22 figures, approx. 15 tables. VI, 362 pages.
ISBN 3-540-96067-8

Contents: Introduction. - Censuses. - Heads or Tails.
Quality Control. - Probabilistic Vocabulary of Measure
Theory. Inventory of the Most Useful Tools. - Inde-
pendence: Statistics Based on the Observation of a
Sample. - Gaussian Samples, Regression, and Analysis
of Variance. - Conditional Expectation, Markov Chains,
Information. - Dominated Statistical Models and
Estimation. - Statistical Decisions. - Bibliography. -
Notation and Conventions. - Index.

D. Dacunha-Castelle, M. Duflo

Probability and Statistics II

Translated by D. McHale

1986. 6 figures. XIV, 410 pages. ISBN 3-540-96213-1

Contents: Introduction to Random Processes. - Time
Series. - Martingales in Discrete Time. - Asymptotic
Statistics. - Markov Chains. - Step by Step Decisions. -
Counting Processes. - Processes in Continuous Time. -
Stochastic Integrals. - Bibliography. - Notations and
Conventions. - Index.

This is a mathematically oriented introduction to statistics, with systematic and serious views of sophisticated
probabilistic techniques which are utilized for the
modernization of concrete situations. The first part of
the work is devoted to descriptive statistics, discrete
probability and random walk, and the second to stochastic processes, statistics of processes, and asymptotic
theories. Together, the book give a modern and up-to-
date view of the theoretical and practical tools of probability and statistics.

Springer-Verlag
Berlin Heidelberg New York
London Paris Tokyo